COASTAL AND
ESTUARINE PROCESSES

ADVANCED SERIES ON OCEAN ENGINEERING

Series Editor-in-Chief
Philip L-F Liu *(Cornell University)*

*The complete list of the published volumes in the series can be found at
http://www.worldscibooks.com/series/asoe_series.shtml

Advanced Series on Ocean Engineering — Volume 29

COASTAL AND ESTUARINE PROCESSES

Peter Nielsen
The University of Queensland, Australia

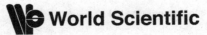 World Scientific

NEW JERSEY · LONDON · SINGAPORE · BEIJING · SHANGHAI · HONG KONG · TAIPEI · CHENNAI

Published by

World Scientific Publishing Co. Pte. Ltd.

5 Toh Tuck Link, Singapore 596224

USA office: 27 Warren Street, Suite 401-402, Hackensack, NJ 07601

UK office: 57 Shelton Street, Covent Garden, London WC2H 9HE

British Library Cataloguing-in-Publication Data
A catalogue record for this book is available from the British Library.

First published 2009
Reprinted 2012

COASTAL AND ESTUARINE PROCESSES
Advanced Series on Ocean Engineering — Vol. 29

Copyright © 2009 by World Scientific Publishing Co. Pte. Ltd.

ISBN-13 978-981-283-711-0
ISBN-10 981-283-711-6
ISBN-13 978-981-283-712-7 (pbk)
ISBN-10 981-283-712-4 (pbk)

Printed in Singapore by B & Jo Enterprise Pte Ltd

Preface

The present text on coastal and estuarine processes is intended for students of civil and/or environmental engineering, and earth or environmental science. It is also hoped that practicing professionals will find it useful as a handbook.

Since wind waves and swell deliver the design forces on coastal structures and are the drivers of most of the morphological processes, they are treated in considerable detail in Chapter 1. That is, the form and magnitudes of velocities and pressures are explored in as much detail as linear wave theory allows. This text emphasizes more than its predecessors the explicit formulation of linear wave theory. That is, most sine wave quantities are expressed in terms of $k_o h$ rather than kh where k is an implicit function of the depth. The improved transparency thus achieved is a clear pedagogical advantage. The explicit expressions have the extra advantage of easy differentiation and integration. A number of new results concerning the growth of forced long waves are included because they bring significant qualitative understanding while being fairly straightforward to derive from linear long wave theory. These new long-wave results significantly enhance the introductory treatment of tsunami, surf beat and storm surges.

The treatment of non-linear waves, including bores, is restricted to a fairly brief summary and a few bibliographical notes.

The treatment of natural waves gives the descriptive and statistical tools, which are necessary for civil- and environmental engineering praxis.

Surf zone processes, the subject of Chapter 2, are covered in enough detail to enable understanding and basic modelling of coastal inundation, longshore currents, undertow and rip currents, using the concept of radiation stress.

Chapter 3 covers wind driven circulation in shallow water and storm surges including linear long waves forced by variations in atmospheric pressure and/or wind shear stress.

The remarkable variability of tidal regimes around the continents is described in Chapter 4 together with the equilibrium theory of tides. Methods of the modelling and prediction of astronomical tides are explained together with the meteorologically driven deviations of real tides from the regular, astronomical ones. Simple analytical descriptions are given for waves in rivers and tides in coastal lagoons and atolls including tidal pumping.

Density driven flows and mixing in estuaries are the subjects of Chapter 5. The pressure distributions in stratified fluids are illustrated and shear instability and mixing at interfaces are treated in terms of force and energy balances. The classical solutions for gradient diffusion are mentioned after which turbulent diffusion is discussed from the Eulerian as well as from the Lagrangian approach. This includes the finite-mixing-length theory, which enables quantification of differential diffusion for different sediment sizes suspended in the same flow.

The basic mechanisms of coastal and estuarine sediment transport are treated in Chapter 6. The emphasis is on non-cohesive sediments. This chapter also includes small and medium scale bedforms and their contribution to the hydraulic roughness. The recent advances in the understanding of acceleration effects and boundary layer streaming on sheetflow sediment transport under waves are included.

Chapter 7 is a brief treatment of the morphology and morphodynamics of beaches, estuaries, tidal inlets and deltas. In addition the management tools of 2D and 3D sediment budgets are discussed.

The importance of waves and tides for the modelling of groundwater dynamics and saltwater intrusion near the coast is illustrated in Chapter 8.

The text is fully indexed and the comprehensive cross-referencing enables it to be used as a handbook as well as a textbook.

Brisbane, April 2008

Peter Nielsen.

Contents

Contents

Contents

Contents

1 Water waves

1.1 INTRODUCTION

1.1.1 The nature of water waves

The sea surface is usually not a flat horizontal surface. Generally it is covered by wind waves, which can have very different periods T depending on how far they have traveled. The little, freshly generated ripples on a pond have periods of the order 0.1 seconds while the long swell waves arriving on a beach from storms a few thousand kilometres away have periods in the range from fifteen to twenty seconds.

Tectonic seabed movements and/or associated landslides can generate tsunami. While these are hardly visible in deep water due to small

Figure 1.1.1: Waves are an important part of the coastal scenery. Usually a source of pleasure but occasionally of great destruction.

height (<1m) and long period 100-2000s, they can cause very serious damage as their runup can inundate vast areas up to 50m above normal sea level.

Observing a beach at six hour time intervals may reveal that some rocks or other fixtures, seem to "come and go". What happens is of course that the mean water surface (MWS) moves up or down due to the tide. The tide is also a wave motion with period about 12 or 24 hours. However, because of its longer period its wavelength L is so large that its curvature is not perceivable. Still longer waves exist in the ocean. Travelling weather systems can generate waves with periods of several days.

In deep water, the wave height is limited through the steepness. Theoretical results for waves propagating with constant form give a limiting steepness of about 14%:

$$(H/L)_{max} \approx 0.14 \qquad (1.1.1)$$

Natural storm waves which change their shape all the time can now and then be much steeper. The largest wave ever measured off Brisbane was measured during Cyclone Roger in March 1993 at a height of 13.2m. This represents the typical extreme wave heights generated by stationary or near-stationary cyclones. Fast moving cyclones may produce waves that are at least twice as big as illustrated in Section 1.6.6. Also, the lining up of several low pressure systems can give rise to very large waves in "the roaring forties" of the southern oceans. Several photographs of extreme waves can be found in Bascom (1980).

Waves that approach a beach will start to break when the depth is roughly equal to the wave height. On a very flat slope, the ratio between wave height and water depth becomes constant and equal to approximately 0.55, cf Nelson (1997). Meeting an opposing current will also cause the waves to change shape and perhaps break.

After breaking due to a bar or a current, new waves with a shorter length and period may emerge.

1.1.2 Wave restoring forces

Waves of different lengths (and periods) are restored or propagated by different physical mechanisms. The shortest ($L \leq 1cm$) waves are dominated by surface tension. Intermediate wavelengths are dominated by gravity and the longest ($T > 24hours$) waves are governed by the Coriolis acceleration. Each of these forces plays a *restoring* role similar to that of the elastic tension in an oscillating rubber band.

1.1.3 Wave generating forces

The most important source of wave generating forces are the travelling weather systems. They can generate wind waves of periods 0.1s to 20s through surface instability under the wind stress on the surface and much longer waves (periods of a few days) through the barometric pressure variations. However, earthquakes, landslides and calving glaciers are other natural wave generators. Tides originate from the Earth's rotation under ocean bulges generated by the non-uniform gravitational pull by the Moon and the Sun. Ships and other watercraft also generate waves, which can be of engineering significance because of bank erosion, particularly in rivers and canals.

1.2 DEFINITIONS AND COORDINATE SYSTEMS

The instantaneous, local water surface elevation is $\eta(x,y,t)$, where x is usually the horizontal coordinate in the direction of wave propagation. The vertical coordinate z is usually measured positive upwards from the mean water surface. The wave height H is the vertical distance from crest to trough. The depth to the mean water surface (MWS) level is called h.

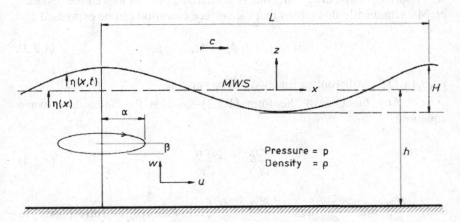

Figure 1.2.1: Definition sketch.

Individual water particles move along paths that are almost closed elliptical orbits with half-axes α and β. The vertical excursion 2β equals the wave height at the surface and vanishes at an impermeable bed.

The horizontal water particle velocity in the direction of wave propagation is $u(x,y,z,t)$ and the vertical particle velocity is $w(x,y,z,t)$. These Eulerian velocities are the most commonly used but the Lagrangian velocities $[u(x_o,y_o,z_o,t), \; v(x_o,y_o,z_o,t), \; w(x_o,y_o,z_o,t)]$ are sometimes more convenient, particularly, for describing the motion above the wave trough.

The local gauge pressure is $p(x,y,z,t)$ and the overpressure compared with the still water situation is called p^+,

$$p(x,y,z,t) = -\rho g z + p^+(x,y,z,t) \tag{1.2.1}$$

1.2.1 Waves that propagate with constant form

If the wave propagates with constant form we can define the speed of propagation c and the wave length is L. There exists the fundamental relation

$$L = cT \tag{1.2.2}$$

between length, speed and period.

Given that the waves propagate at speed c with a constant form, the water surface shape after t seconds is unchanged but has moved the distance ct. Mathematically this means that the surface elevation can be expressed as

$$\eta(x,t) = f(\phi) = f(x - ct) \tag{1.2.3}$$

i e, $\eta(x,t)$ is really only a function of one variable.

Any function of the form (1.2.3) satisfies the linear long wave equation:

$$\frac{\partial^2 \eta}{\partial t^2} = c^2 \frac{\partial^2 \eta}{\partial x^2} \tag{1.2.4}$$

since $\dfrac{\partial^2 \eta}{\partial t^2} = c^2 \dfrac{\partial^2 f}{\partial \phi^2}$ and $\dfrac{\partial^2 \eta}{\partial x^2} = \dfrac{\partial^2 f}{\partial \phi^2}$. Positive c corresponds to propagation in the positive x-direction and vice versa.

Waves that propagate with constant form also offer the option of working in a frame of reference that folows the waves. In this frame of reference, the water motion is steady. Hence the term "steady waves" is also sometimes used for waves that propagate with constant form.

1.2.2   Exercise

Consider the function

$$\eta(x,t) = \frac{H}{2}\cos(\frac{2\pi}{L}x - \frac{2\pi}{T}t) = \frac{H}{2}\cos(kx - \omega t) \qquad (1.2.5)$$

which represents the surface elevation of a sine wave with height H, period T and length L. It has the form (1.2.3) since it can also be written

$$\eta(x,t) = \frac{H}{2}\cos[\frac{2\pi}{L}(x - \frac{L}{T}t)] = \frac{H}{2}\cos[\frac{2\pi}{L}(x - ct)] \qquad (1.2.6)$$

Write a spreadsheet program which enables you to choose different values of H, L and T and to plot snapshots of η for fixed times, t_i, i e, $\eta(x,t_i)$ and time series of η at various points x_j, i e, $\eta(x_j,t)$. Note, in particular, how $\eta(x,t_i)$ moves along the x-axis with time.

It follows from the continuity principle alone, that if a wave propagates with constant form, the flow rate must vary in step with the surface elevation.

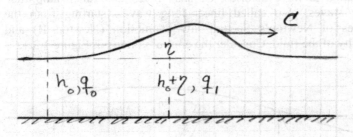

Figure 1.2.2.

Consider two cross sections in Figure 1.2.2. One with depth h_o and flow rate q_o the other with depth $h_1 = h_o + \eta$ and flow rate q_1. In the frame of reference which follows the wave, the scenario is steady, so the surface between the two cross sections does not move up or down. This requires "outflow = inflow", i e,

$$q_1 - (h_o + \eta)c = q_o - h_o c \qquad (1.2.7)$$

or

$$q_1 - q_o = c\eta \qquad (1.2.8)$$

Hence, for a steady wave the flow rate varies in step with the surface elevation, and the "exchange rate" $dq/d\eta$ is c.

1.3 SMALL AMPLITUDE SHALLOW WATER WAVES

The *shallow water* or *long wave* approximation: $h/L \ll 1$ often leads to workable analytical solutions, which are valuable first steps towards an understanding of coastal wave phenomena. This section is therefore dedicated to such waves.

In one horizontal dimension it is possible to obtain simple analytical solutions for forced as well as free waves. Most of these solutions have been known for a long time, but Section 1.3.3 contains new insights into the transient stages, e g, the growth process of forced long waves obtained by considering the developing waves as superpositions of forced and free waves.

In two horizontal dimensions, much fewer simple analytical solutions are available. However, in Section 1.3.2, we shall consider coastal trapped long waves, i e, long waves which hug the shoreline rather than sending their energy off-shore, because these provide important insights into the nature of some storm surges.

1.3.1 The 1D shallow water wave equation

As an introduction to wave theories, we shall investigate small amplitude shallow water waves, also called linear long waves, starting from the basic principles of Newton's Second law of motion (Newton II) and incompressibility of the fluid. Consider the control volume in Figure 1.3.1.

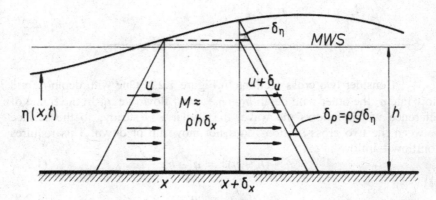

Figure 1.3.1: In long wave- or shallow-water wave theory, p is assumed hydrostatic and u is uniform throughout the depth. The fluid in the control volume will be accelerated due to the difference in hydrostatic pressure force and, if the in-flow is greater than the out-flow, the surface must go up.

Assuming small amplitude, $H \ll h$, and vertically uniform velocity, mass times acceleration for the enclosed water is $\rho h \delta_x (\frac{\partial u}{\partial t} + u \frac{\partial u}{\partial x}) \approx$ $\rho h \delta_x \frac{\partial u}{\partial t}$ and the horizontal pressure force due to the surface elevation increment δ_η is approximately $-\rho g \delta_\eta h$. Hence,

Newton II:
$$\rho h \delta_x \frac{\partial u}{\partial t} = -\rho g \delta_\eta h \qquad (1.3.1)$$

or

$$\frac{\partial u}{\partial t} = -g \frac{\partial \eta}{\partial x} \qquad (1.3.2)$$

Incompressibility requires the surface to go down if the outflow $(h + \eta + \delta_\eta)(u + \delta_u)$ is greater than the inflow $(h + \eta)u$. Hence,

Continuity:
$$\frac{\partial \eta}{\partial t} = -\frac{\partial([h + \eta]u)}{\partial x} \approx -h \frac{\partial u}{\partial x} \qquad (1.3.3)$$

Elimination of u between the two equations leads to the linear shallow water wave equation

$$\frac{\partial^2 \eta}{\partial t^2} = gh \frac{\partial^2 \eta}{\partial x^2} \qquad (1.3.4)$$

which, by analogy with the general wave equation (1.2.4), shows that all wave shapes are possible in shallow water and that the speed of propagation for small amplitude shallow water waves is

$$c = \sqrt{gh} \qquad (1.3.5)$$

1.3.1.1 Shallow water sine waves
As discussed in Section 1.3.1 the linear wave equation (1.3.4) allows any wave shape $f(x \pm \sqrt{gh}\,t)$. However, waves of sinusoidal shape with period T, length L and amplitude $H/2$ are of special interest:

$$\eta(x,t) = \frac{H}{2} \cos \frac{2\pi}{L}(x - ct) = \frac{H}{2} \cos(\frac{2\pi}{L}x - \frac{2\pi}{T}t) = \frac{H}{2} \cos(kx - \omega t) \quad (1.3.6)$$

$k = \dfrac{2\pi}{L}$ is called the wave number and $\omega = \dfrac{2\pi}{T}$ the angular frequency.

The velocity field is found by applying the linearised continuity principle, which for the control volume in Figure 1.3.1 gives

$$\frac{\partial \eta}{\partial t} = -h \frac{\partial u}{\partial x} \qquad (1.3.3)$$

This leads to

$$u(x,t) = \frac{H}{2h} \frac{\omega}{k} \cos(kx - \omega t) + \text{constant} \qquad (1.3.7)$$

and since

$$\frac{\omega}{k} = \frac{2\pi / T}{2\pi / L} = \frac{L}{T} = c = \sqrt{gh} \qquad (1.3.8)$$

we get

$$u(x,t) = \frac{H}{2h} \sqrt{gh} \cos(kx - \omega t) + \text{constant} \qquad (1.3.9)$$

which shows that $u(x,t) = \dfrac{\eta(x,t)}{h} \sqrt{gh}$ for a shallow water wave. In particular, this means that maximum horizontal velocity coincides with the passage of the wave crest. Note that this in-step variation of u and η occurs only for waves which propagate with constant form. For a decaying wave, u peaks before η as discussed in relation to river tides in Section 4.5.

The arbitrary constant of integration means that the wave motion can be superimposed on any uniform current without violating the governing equations (1.3.2) and (1.3.3).

The vertical velocity can subsequently be found by applying continuity locally, i e,

$$\frac{\partial u}{\partial x} + \frac{\partial w}{\partial z} = 0 \qquad (1.3.10)$$

and noting that the vertical velocity must vanish at the impermeable bed, $w(x,-h,t) \equiv 0$. This leads to

$$w(x,z,t) = \int_{z=-h}^{z} -\frac{\partial u}{\partial x} dz = \int_{-h}^{z} \frac{H}{2h} \sqrt{gh}\, k \sin(kx - ct)\, dz \qquad (1.3.11)$$

where we invoke (1.3.8) in the form $\sqrt{gh}\,k = \omega$ and get

$$w(x,z,t) = \omega\frac{H}{2}\frac{z+h}{h}\sin(kx - \omega t) \tag{1.3.12}$$

For shallow water waves, it is sometimes convenient to use a vertical coordinate z' with origin at the bottom instead of the surface. The expression for the vertical velocity then becomes

$$w(x,z',t) = \omega\frac{H}{2}\frac{z'}{h}\sin(kx - \omega t) \tag{1.3.12}$$

showing that, while $u(x,t)$ varies in step with $\eta(x,t)$, $w(x,t)$ varies in step with $\dfrac{\partial \eta}{\partial t}$, i e, $w(x,t) = \dfrac{z'}{h}\dfrac{\partial \eta}{\partial t}$. As for u and η, the timing relation between w and η is different if the waves are not propagating with constant form.

1.3.2 Long waves in two horizontal dimensions

In two horizontal dimensions the linearised momentum and continuity equations (1.3.2) and (1.3.4) are generalised to

Momentum:
$$\begin{aligned}\frac{\partial u}{\partial t} &= -g\frac{\partial \eta}{\partial x} \\ \frac{\partial v}{\partial t} &= -g\frac{\partial \eta}{\partial y}\end{aligned} \tag{1.3.13}$$

and

Continuity:
$$\frac{\partial \eta}{\partial t} \approx -h\left(\frac{\partial u}{\partial x} + \frac{\partial v}{\partial y}\right) \tag{1.3.14}$$

In analogy with the 1D equation, these equations allow plane gravity waves of arbitrary shape to propagate with constant form in any direction with speed $c = \sqrt{gh}$, i e,

$$\eta(x,y,t) = A f(x \pm c_x t, y \pm c_y t) = A f_2(k_x x \pm \omega t, k_y\, y \pm \omega t) \tag{1.3.15}$$

9

where $c_x^2 + c_y^2 = gh$ and $k_x^2 + k_y^2 = k^2 = \left(\dfrac{2\pi}{\sqrt{gh}\,T}\right)^2$, or with $\vec{r}=(x,y)$ and

$\vec{k}=(k_x,k_y)$

$$\eta(\vec{r},t) = A\mathrm{f}_1(\vec{k}\cdot\vec{r} \pm \sqrt{gh}\,t) \qquad (1.3.16)$$

However, since one of the most important applications concerns storm surges, which often have periods longer than one day, and hence are influenced by the Earth's rotation, we introduce the Coriolis effect via the Coriolis parameter $f_c = 2\Omega\sin\varphi$, where $\Omega = 7.27\times10^{-5}$rad/s is the Earth's angular frequency and φ is the latitude ($0<\varphi<\pi/2$ in the northern hemisphere and $-\pi/2<\varphi<0$ in the southern hemisphere).

In a coordinate system which rotates with the Earth the Coriolis effect is an apparent push to the right in the northern hemisphere and to the left in the southern hemisphere, which modifies the momentum equations (1.3.13) to

$$\frac{\partial u}{\partial t} = -g\frac{\partial\eta}{\partial x} + f_c v$$

$$(1.3.17)$$

$$\frac{\partial v}{\partial t} = -g\frac{\partial\eta}{\partial y} - f_c u$$

The "Coriolis terms" quantify the waves preference for a straight trajectory in the absolute, inertial frame of reference, which corresponds to curved parths in the rotating Earth frame of reference, or to make the waves hug the coastlines of continents while travelling around them clockwise in the northern hemisphere and anticlockwise in the southern hemisphere.

1.3.2.1 Coastal trapped waves

Kelvin waves are long ($T \geq 24$hours) coastal trapped waves. That is, they show the above mentioned hugging of the shorelines of continents.

For the simplified scenario with constant ocean depth they can be described as follows: Imagine a wave which propagates in the x-direction along an infinitely long vertical wall at $y=0$, and with velocities and accelerations in the y-direction which are negligile, see Figure 1.3.2.

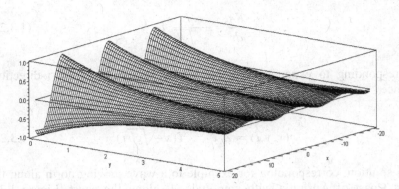

Figure 1.3.2: Coastal trapped wave with amplitude which decays exponentially away from the coastline, y=0. Kelvin waves can be trapped by the Coriolis effect along a vertical coast but only propagate in one direction. Pure gravity waves can be trapped by refraction along a sloping coast and may propagate in both directions, thus having the potential of forming standing wave patterns.

The second part of (1.3.17) then says that, due to the Coriolis effect, there is an equilibrium surface slope given by

$$\frac{\partial \eta}{\partial y} = -\frac{f_c u}{g} \qquad (1.3.18)$$

i e, a slope of magnitude $\dfrac{|f_c|u}{g}$ upward to the right in the northern hemisphere, and upward to the left in the southern hemisphere.

Assuming that the wave propagates with constant form in the x-direction, it can then be described by

$$\eta(x,y,t) = A(y)f(x-\sqrt{gh}\,t) \qquad (1.3.19)$$

and the corresponding horizontal velocity is given by

$$u = \frac{\eta}{h}\sqrt{gh} = A(y)\sqrt{\frac{g}{h}}f(x-\sqrt{gh}\,t) \qquad (1.3.20)$$

Equation (1.3.18) then gives

11

$$\frac{dA}{dy} = -\frac{f_c}{\sqrt{gh}}A \qquad (1.3.21)$$

corresponding to exponential decay of the amplitude in the y-direction. Hence,

$$\eta(x,y,t) = A_o\, e^{-\frac{f_c}{\sqrt{gh}}y}\, f(x - \sqrt{gh}\,t) \qquad (1.3.22)$$

is a solution, corresponding for example to a wave moving north along the East Coast of Australia with amplitude A_0 along the coast. Figure 1.3.3 shows Kelvin waves, generated by weather systems in the Indian Ocean travelling anticlockwise around western and southern Australia.

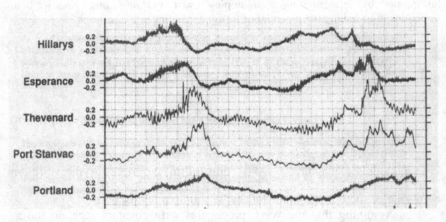

Figure 1.3.3: Tidal anomalies (actual tide – astronomical tide) travelling anticlockwise around the western and southern coast of Australia. The horizontal time axis covers the month of June 1995. The verical axis is in metres. The fact that it takes the surges about one day to travel from Esperance to Thevenard, ca 1100km, corresponds to a speed of 13m/s and through long wave theory to an equivalent constant depth around 16m. Data from The Australian National Tide Facility, Annual Report (1995).

Kelvin waves cannot propagate in the opposite direction along the same coast as their energy is quickly radiated offshore. If the ocean depth is not constant, these features of the Kelvin waves are qualitatively unchanged but details are different, e g, the shorenormal y-dependence of the amplitude is no longer exponential.

We note that if the ocean depth is not constant, observed surges may include coastal trapped waves of a different nature: Stokes (1846) realised that along a linearly sloping coast, say $h = y \tan\beta$, trapped gravity waves may propagate in both directions with speed $c = \sqrt{\dfrac{gL}{2\pi}}\sqrt{\tan\beta} = \sqrt{\dfrac{g}{k}}\sqrt{\tan\beta}$.

These, so-called *edge waves*, also have exponentially decaying amplitude, $A(y) \sim e^{-ky}$.

Later work by Ursell (1952) showed that Stokes' solution was in fact only the first of infinitely many modes and that coastal trapping of gravity waves is not restricted to shallow water waves. For these shorter waves, typical periods one to five minutes, the tendency to hug the shoreline is due to refraction.

Also note that since the trapped gravity waves can propagate in both directions, standing edge waves may occur for example near a headland. Such standing edge waves with lengths in the order tens to hundreds of metres have been suggested as forcing mechanisms for morphological features like cusps and rip cells. However, the "chicken and egg question" on this issue is still unresolved.

1.3.3 Forced shallow water waves

1.3.3.1 Equations of motion

We saw in the previous section that the linearised momentum equation (1.3.2) combines with the small amplitude continuity equation (1.3.3) to give the linear wave equation (1.3.4) and that this equation allows *free waves* of any shape to propagate at speed $\pm\sqrt{gh}$ without change of shape.

We shall now investigate the shallow water wave motions which are generated if a forcing, e g, wind stress or surface pressure, acts on the sea surface. We derive the general wave equation for long waves and solve it for a few cases of forcing with the form f(x-ct), i e, forcing which travels with constant form and a velocity c, which is usually different from the free wave celerity $\sqrt{gh}$.

For most types of forcing the ocean response in the steady state is a forced wave with the same shape as the forcing, i e, $\eta_{forced}(x,t) = A_{forced}f(x$-$ct)$, where the amplitude A_{forced} depends on the strength of the forcing and its relative speed $c/\sqrt{gh}$.

We shall also determine the particular set of free waves which are required in order to satisfy the condition of flat surface and no water motion at the onset of a henceforth steady forcing. The development of the wave form from the initial flat surface towards the asymptotic steady solution, is then a result of the free waves moving away from the forcing. The developing wave usually appears highly unsteady, including shape changes, although it is a simple superposition of steady waves.

These linear, superposition-based solutions show extensive analogy with the linear solutions for a simple forced oscillator, see e g Kreyszig (2006) Section 2.8.

In the resonant case of $c = \sqrt{gh}$, the linear equations have no steady solution. Instead they predict a linearly growing wave of the form $-t\sqrt{gh}\dfrac{\partial}{\partial x}f(x-ct)$.

The new wave equations, which include forcing, are derived on the basis of Figure 1.3.4, which is Figure 1.3.1 with various types of forcing added.

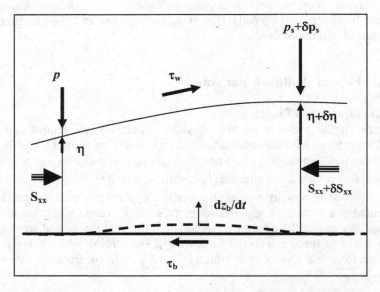

Figure 1.3.4: Surface waves may be forced by variable atmospheric pressure p_s, wind shear stress τ_w, and variable radiation stress S_{xx} from short waves. In addition, sea bed movements dz_b/dt may generate tsunami, and the waves may be losing energy to bed friction, τ_b.

Newton II applied to this control volume gives

$$\frac{\partial u}{\partial t} + u\frac{\partial u}{\partial x} = -g\frac{\partial \eta}{\partial x} - \frac{1}{\rho}\frac{\partial p_s}{\partial x} - \frac{1}{\rho h}\frac{\partial S_{xx}}{\partial x} + \frac{\tau_w}{\rho h} - \frac{\tau_b}{\rho h} \tag{1.3.23}$$

which for small amplitude waves, $\eta \ll h$, becomes

$$\frac{\partial u}{\partial t} = -g\frac{\partial \eta}{\partial x} - \frac{1}{\rho}\frac{\partial p_s}{\partial x} - \frac{1}{\rho h}\frac{\partial S_{xx}}{\partial x} + \frac{\tau_w}{\rho h} - \frac{\tau_b}{\rho h} \tag{1.3.24}$$

The continuity equation, simplified by assuming $\eta \ll h$ and that the bed deformation varies slowly in the x-direction: $\frac{h^2}{2}\frac{\partial^2}{\partial x^2}\frac{\partial z_b}{\partial t} \ll \frac{\partial z_b}{\partial t}$, reads:

$$\frac{\partial \eta}{\partial t} = -h\frac{\partial u}{\partial x} + \frac{\partial z_b}{\partial t} \tag{1.3.25}$$

As for the free wave case, the linear wave equation for forced long waves is obtained by eliminating u between these two equations. We get

$$\frac{\partial^2 \eta}{\partial t^2} = gh\frac{\partial^2 \eta}{\partial x^2} + \frac{1}{\rho}h\frac{\partial^2 p_s}{\partial x^2} + \frac{1}{\rho}\frac{\partial^2 S_{xx}}{\partial x^2} - \frac{1}{\rho}\frac{\partial \tau_w}{\partial x} + \frac{1}{\rho}\frac{\partial \tau_b}{\partial x} + \frac{\partial^2 z_b}{\partial t^2} \tag{1.3.26}$$

We note the factor h present in the pressure forcing term but not in the S_{xx} term or the wind forcing term, meaning that pressure forcing (a volume force) is the more efficient in deeper water.

This equation being linear allows for the response to each kind of forcing to be evaluated seperately and subsequently superposed.

If the bottom displacement is rapid, the assumption of hydrostatic pressure, corresponding to $\frac{\partial^2 z_b}{\partial t^2} \ll g$ is violated, as in some earthquake scenarios. The equation then becomes

$$\frac{\partial^2 \eta}{\partial t^2} = (g + \frac{\partial^2 z_b}{\partial t^2})h\frac{\partial^2 \eta}{\partial x^2} + \frac{1}{\rho}h\frac{\partial^2 p_s}{\partial x^2} + \frac{1}{\rho}\frac{\partial^2 S_{xx}}{\partial x^2} - \frac{1}{\rho}\frac{\partial \tau_w}{\partial x} + \frac{1}{\rho}\frac{\partial \tau_b}{\partial x} + \frac{\partial^2 z_b}{\partial t^2} \tag{1.3.27}$$

1.3.3.2 Steady forced solutions

It turns out that Equation (1.3.26), like the unforced linear shallow water equation, has steady solutions of arbitrary shape if bottom friction is negligible.

That is, for p_s, S_{xx} or z_b forcing of the form $F_o \times f(x-ct)$, there are steady solutions of the form $\eta_{forced}(x,t) = A_{forced} \times f(x-ct)$.

For wind stress, which appears in terms a 1st rather than 2nd derivative in (1.3.26) the result is slightly different: Wind stress forcing in the form, $\tau_o \times f(x-ct)$ generates $\eta_{forced}(x,t) = A_{forced} \times F(x-ct)$, where $f = F'$.

First consider forcing from variable surface pressure: $p_s(x,t) = P_o \times f(x-ct)$ only. Inserting this expression for $p_s(x,t)$ into (1.3.26) with $\tau_b = \tau_w = S_{xx} = \dfrac{\partial z_b}{\partial t} = 0$, and assuming the form $\eta_{forced}(x,t) = A_{forced} \times f(x-ct)$ for the solution, immediately gives

$$c^2 A_{forced}\, f'' = gh\, A_{forced}\, f'' + \frac{P_o}{\rho} h f'' \qquad (1.3.28)$$

and hence the no-friction solution

$$A_{forced} = \frac{-P_o/\rho g}{1-c^2/gh} \qquad (1.3.29)$$

which is valid for any small amplitude wave shape. This result may alternatively be derived using steady flow considereations as done by Dean and Dalrymple (1989). For slow moving pressure systems, $c^2 < gh$, we see that the response is negative, i e, a low pressure system generates a positive surge. A fast moving moving, $c^2 > gh$, low pressure system however generates a depression of the water surface. This last, somewhat paradoxical result is discussed further in Section 3.4 which gives more details about pressure driven storm surges.

For waves driven by a steadily moving bottom deformation $z_b = Z_o \times f(x-ct)$ the solution is similar, but it must be a little different, since common sense requires it to vanish for $c \to 0$. Stationary bumps on the bottom of a bath tub do not deform the surface. In this case, insertion of the forcing form into the wave equation (1.3.26) gives

$$c^2 A_{forced}\, f'' = gh\, A_{forced}\, f'' + c^2 Z_o\, f'' \qquad (1.3.30)$$

and hence

$$A_{\text{forced}} = \frac{c^2 Z_o}{c^2 - gh} \qquad (1.3.31)$$

The solution for τ_w-forcing alone is again similar but different in that τ_w appears in (1.3.26) via $\dfrac{\partial \tau_b}{\partial x}$ while p_s is represented by $\dfrac{\partial^2 p_s}{\partial x^2}$. A solution, analogous with (1.3.29), to the wind stress problem can therefore be expressed in terms of $T_w(x,t) = T_o \times f(x-ct)$, where $\tau_w = \dfrac{\partial T_w}{\partial x}$. Insertion into (1.3.26) then gives

$$c^2 A_{\text{forced}} \, f'' = gh \, A_{\text{forced}} \, f'' - \frac{T_o}{\rho} f'' \qquad (1.3.32)$$

and the solution, which is illustrated by Figure 1.3.5, is $\eta_{\text{forced}}(x,t) = A_{\text{forced}} \times f(x-ct)$ with

$$A_{\text{forced}} = \frac{T_o / \rho gh}{1 - c^2 / gh} \qquad (1.3.33)$$

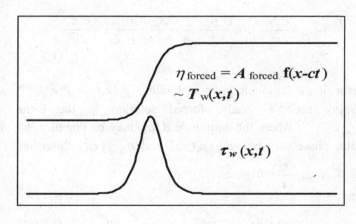

$$\eta_{\text{forced}} = A_{\text{forced}} \, f(x\text{-}ct)$$
$$\sim T_w(x,t)$$

$$\tau_w(x,t)$$

Figure 1.3.5: A hump shaped $\tau_w(x,t)$ moving with constant form generates a step shaped surge in the steady state.

1.3.3.3 Influence of bed friction on steady forced waves

We saw above, that the friction free solutions are (except for wind stress forcing) proportional to the forcing. The factor of proportionality being positive or negative can then be interpreted as the forced wave being precisely in phase respectively 180° out of phase with the forcing. The effect of linearised friction on sine waves is a phase shift ($\neq$ 0°, 180°), which enables the forcing to do work on the water surface in order to compensate for the frictional energy loss. We derive an expression for the frictional phase shift of a steady simple harmonic wave, which shows complete analogy with a simple forced 1D oscillator.

For steady simple harmonic waves, we linearise the non-linear friction term $\tau_b = \frac{1}{2}\rho f_w |u| u$ using the Fourier expansion

$$|u|u = |U\sin\omega t| U\sin\omega t = \frac{8U^2}{3\pi}\sin\omega t - \frac{8U^2}{15\pi}\sin 3\omega t + \dots$$

(1.3.34)

and introduce $D = 4f_w U/3\pi h$, which is a constant for steady waves only, to get the linear approximation

$$\tau_b \approx \frac{1}{2}\rho f_w \frac{8U}{3\pi} u = \rho h D u$$

(1.3.35)

with which the linearised wave equation (1.3.26) becomes

$$\frac{\partial^2 \eta}{\partial t^2} + D\frac{\partial \eta}{\partial t} = gh\frac{\partial^2 \eta}{\partial x^2} + \frac{1}{\rho}h\frac{\partial^2 P_s}{\partial x^2} + \frac{1}{\rho}\frac{\partial^2 S_{xx}}{\partial x^2} - \frac{1}{\rho}\frac{\partial \tau_w}{\partial x} + \frac{\partial^2 z_b}{\partial t^2}$$

(1.3.36)

We then consider a simple harmonic pressure forcing: $p_s(x,t) = P_o e^{i(\omega t - kx)}$ and correspondingly seek a steady, forced solution in the form $\eta_{forced}(x,t) = A_{forced} e^{i(\omega t - kx)}$, where the amplitude A_{forced} may be complex, to allow for a possible phase shift between $p_s(x,t)$ and $\eta_{forced}(x,t)$. Insertion into (1.3.36) with S_{xx}, τ_w, $\frac{\partial z_b}{\partial t} \equiv 0$ gives:

$$-\omega^2 A_{forced} + i\omega D A_{forced} = -k^2 gh A_{forced} - \frac{k^2}{\rho}h P_o$$

(1.3.37)

and hence, with $k^2 gh = \omega_o^2$ and $k^2 c^2 = \omega^2$:

18

$$A_{\text{forced}} = \frac{-P_o / \rho g}{1-(\frac{\omega}{\omega_o})^2 + i\frac{\omega D}{\omega_o^2}} = \frac{-P_o / \rho g}{1-\frac{c^2}{gh} + i\frac{cD/k}{gh}} \qquad (1.3.38)$$

which shows analogy with the solution of the ordinary differential equation that represents the behaviour of a simple forced oscillator with linear damping. This analogy is due to the fact that functions describing waves with constant form are functions of a single variable, namely $x-ct$.

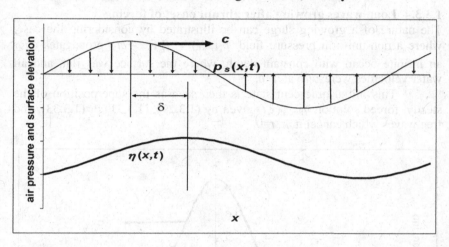

Figure 1.3.6: The energy transfer between the forcing, in this case the air pressure p_s, and the water surface depends on the relative phase. For $0<\delta<L/2$ the wave crest moves towards lower pressure, so the pressure field is doing positive work on the wave. For simple harmonic shapes, the energy transfer rate is proportional to $\sin k\delta$ and hence maximum for $\delta=L/4$.

The solution corresponding to (1.3.38) may be written as

$$\eta(x,t) = |A_{\text{forced}}|\cos(\omega t - k[x-\delta]) \qquad (1.3.39)$$

where the distance between the pressure peak and the water surface peak is

$$\delta = \frac{1}{k}\tan^{-1}(\frac{\omega D / \omega_o^2}{1-(\omega/\omega_o)^2}) \qquad (1.3.40)$$

see Figure 1.3.6.

19

The steady solutions above cover the asymptotic case, where there is no longer any growth. For the friction-free case, zero growth corresponds to $\delta = 0$ or $\delta = L/2$ in the notation of Figure 1.3.6. For $\delta = 0, L/2$, the pressure field does no work on the moving water surface. In the steady state with friction, $\delta \neq 0, L/2$, there is balance between the work done by the pressure field and the dissipation due to the bottom shear stress. Analogous results are found for simple harmonic wind and S_{xx} forcing.

1.3.3.4 Long waves growing after abrupt onset of forcing

The nature of a growing surge can be illustrated by considering the case, where a non-uniform pressure field $p_s(x,t) = P_o f(x - ct)$ propagates over an infinite ocean with constant depth, where the surface was flat, and all water velocities were zero at $t=0$.

This initial quiescent state is then seen as the superposition of the steady, forced solution $\eta_{\text{forced}}(x,t)$ given by (1.3.29), (1.3.31) or (1.3.33), and free waves which cancel it at $t=0$.

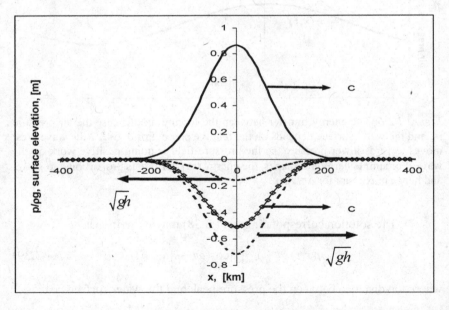

Figure 1.3.7: The initial flat surface is seen as the superposition of the steady, forced solution (solid line) and two free waves given by (1.3.43) which cancel it with respect to surface elevations and water velocities at $t=0$.

20

The steady solution moves with the air pressure field (◊) at speed c, while the free waves move with speed $\sqrt{gh}$ in opposite directions.

In general, two free waves are required, a right moving free wave $\eta^+_{\text{free}}(x,t) = A^+_{\text{free}} f(x - \sqrt{gh}\, t)$ and a left moving free wave $\eta^-_{\text{free}}(x,t) = A^-_{\text{free}} f(x + \sqrt{gh}\, t)$, so the total surface elevation is:

$$\eta(x,t) = \eta_{\text{forced}}(x,t) + \eta^+_{\text{free}}(x,t) + \eta^-_{\text{free}}(x,t)$$

$$= A_{\text{forced}} f(x - ct) + A^+_{\text{free}} f(x - \sqrt{gh}\, t) + A^-_{\text{free}} f(x + \sqrt{gh}\, t) \qquad (1.3.41)$$

A growing wave emerges as the free waves separate from $\eta_{\text{forced}}(x,t)$ due to their different speed and perhaps decay due to friction, see Figures 1.3.7 and 1.3.8 for the case of surface pressure forcing.

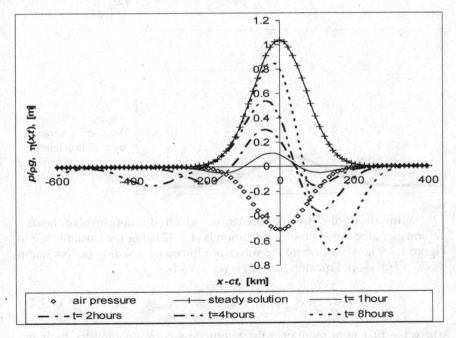

Figure 1.3.8: Developing storm surge driven by a low pressure system (◊ ~ $p/\rho g$) switched on over a calm sea, at $t=0$ and moving at $c=10$m/s while $\sqrt{gh}=14$m/s. $P_o/\rho g= -0.5$m. $A_{\text{forced}}=1.04$m, $(A^+_{\text{free}}, A^-_{\text{free}}) = (-0.89$m, -0.15m$)$. The total $\eta(x,ct)$ is highly unsteady although it is a simple superposition of steady waves.

The relative sizes of the three waves depend on the speed of the forcing as follows: All of these waves being shallow water waves with the same shape, the requirement of zero velocity at $t=0$ leads to

$$A^+_{\text{free}}\sqrt{gh} - A^-_{\text{free}}\sqrt{gh} + A_{\text{forced}}c = 0 \qquad (1.3.42)$$

for all forcing scenarios without bottom deformation.

This is exact if all of the waves propagate with constant form, i e, no friction, so that the velocity is simply $u=(\eta/h)\times$wavespeed.

Together with the flat surface condition $A^+_{\text{free}} + A^-_{\text{free}} + A_{\text{forced}} = 0$ Equation (1.3.42) gives

$$(A^+_{\text{free}}, A^-_{\text{free}}) = \left(-\frac{A_{\text{forced}}}{2}[1+\frac{c}{\sqrt{gh}}], \ -\frac{A_{\text{forced}}}{2}[1-\frac{c}{\sqrt{gh}}]\right) \qquad (1.3.43)$$

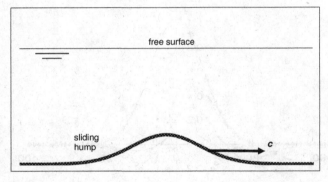

free surface

sliding hump

c

Figure 1.3.9: Tsunami generation by a sliding hump.

In all of the forcing scenarios, which do not involve bottom deformation, the initial no flow condition is (1.3.42). For the tsunami case in Figure 1.3.9 however, where the waves are forced by a sliding bottom hump $z_b(x,t) = H_0 f(x-ct)$, Equation (1.3.42) is replaced by

$$-H_o c + A^+_{\text{free}}\sqrt{gh} - A^-_{\text{free}}\sqrt{gh} + A_{\text{forced}}c = 0 \qquad (1.3.44)$$

where the first term represents the return flow corresponding to the hump moving under a flat surface. With A_{forced} given by (1.3.31), this leads to the friction free result in Figure 1.3.10 for the free waves associated with a simple tsunami:

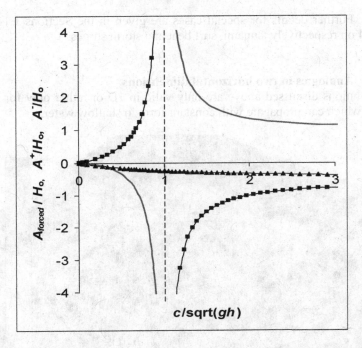

Figure 1.3.10: Relative amplitudes of the steady solution A_{forced}, solid line. The forward moving free wave A_{free}^{+} ■, and of the backward moving free wave A_{free}^{-} ▲ for long waves generated by a hump of height H_o sliding along a horizontal seabed, i e, a simple tsunami.

$$\left(\frac{A_{free}^{+}}{H_o}, \frac{A_{free}^{-}}{H_o}\right) = \left(\frac{1}{2}\frac{c}{\sqrt{gh}-c}, -\frac{1}{2}\frac{c}{\sqrt{gh}+c}\right) \qquad (1.3.45)$$

1.3.3.5 Special behaviour with resonant and near-resonant forcing

For all types of forcing the steady, linear solution degenerates for $c = \sqrt{gh}$ as $A_{forced}, A_{free}^{+} \to \infty$ for $c \to \sqrt{gh}$. What happens then is that η_{forced} and η^{+} merge into a linearly growing wave of the form $t\sqrt{gh}\frac{\partial}{\partial x}f(x-\sqrt{gh}\,t)$. For example, a steady Gaussian-shaped pressure forcing at resonance, drives an N-wave which grows linearly with time.

Further details for special cases are given in the Sections 1.11, 2.3 and 3.4 on respectively tsunami, surf beat and storm surges.

1.3.3.6 Analogies in two horizontal dimensions
The solutions discussed above are only valid in 1D or rather only for plane waves which can propagate with constant form in shallow water.

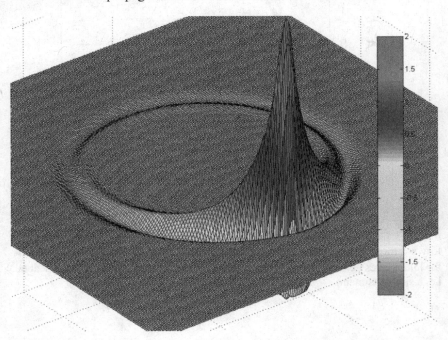

Figure 1.3.11: Resonant 2DH storm surge generated by a Gaussian shaped low pressure system, $p_s(x,y,t) = P_o e^{-[(\frac{x-\sqrt{gh}t}{L})^2 + (\frac{y}{L})^2]}$ propagating from the upper left towards the lower right. The main feature is an N-shaped surge centred on the weather system as detailed in Figure 1.3.12.

Consequently some qualitative differences exist between these and surges in two horizontal dimensions:
- Waves which spread from a central forcing region in 2D must decrease in height $H \sim 1/\sqrt{R}$ in order that the total outward energy flux can be the same through circles of any radius around the source.

- The enhancement of the 1D solutions $\sim gh/(gh-c^2)$ only relates to the direction of propagation while energy corresponding to any excess above a stationary solution ($c=0$) is radiated in the perpendicular direction.

Still, the 1D solutions do give understanding of several features of the 2D solutions. They give the scaling and basis for understanding the intricate shape developments in the direction of forcing propagation, say along $y=0$. In particular, the form $-t\sqrt{gh}\dfrac{\partial}{\partial x}f(x-ct)$ along $y=0$, of the resonant solution is also generally recognised in numerical 2D resonant solutions like the one in Figure 1.3.11.

Differences between $\eta(x,0,t)$ and 1D solutions for surface shear stress forcing are discussed in relation to storm surges in Section 3.3.2.

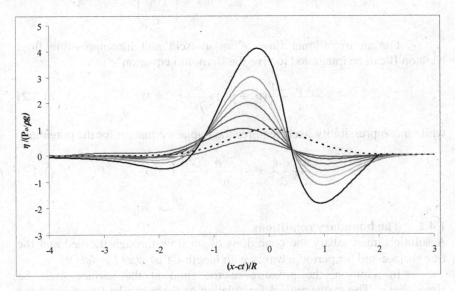

Figure 1.3.12: Cross section in the direction of propagation of the 2D resonant surge in Figure 1.3.11. The dotted curve is the inverted low pressure forcing: $-(P_o/\rho g)$ f($x-ct$) along $y=0$, while the full lines show sections through the developing surge at successive times. Calculations by D P Callaghan using Mike21.

1.4   SINE WAVES IN ARBITRARY DEPTH

1.4.1   The field equations

To get a simple wave theory for arbitrary, i e, not necessarily shallow water depth, we need to relax the previous assumption of vertically uniform velocity and hydrostatic pressure. Still, the underlying principles are the same: <u>Newton II</u> and <u>Continuity.</u>

For mathematical convenience a two-dimensional (*xz*) solution is sought in the form of the velocity potential ϕ which can be defined for an irrotational flow by

$$\vec{u} = -\overline{grad}\,\phi \iff \begin{pmatrix} u \\ w \end{pmatrix} = \begin{pmatrix} -\dfrac{\partial \phi}{\partial x} \\ -\dfrac{\partial \phi}{\partial z} \end{pmatrix} \tag{1.4.1}$$

For an irrotational flow of an inviscid and incompressible fluid Newton II can be integrated to give the Bernoulli equation

$$-\frac{\partial \phi}{\partial t} + \frac{1}{2}(u^2 + w^2) + \frac{p}{\rho} + gz = 0 \tag{1.4.2}$$

while incompressibility itself requires the Laplace equation for the potential

$$\frac{\partial u}{\partial x} + \frac{\partial w}{\partial z} = -\left(\frac{\partial^2 \phi}{\partial x^2} + \frac{\partial^2 \phi}{\partial z^2}\right) \equiv 0 \tag{1.4.3}$$

1.4.2   The boundary conditions

A solution must satisfy the conditions of no flow through the bed and the free surface and for periodic waves with length *L* that $\phi(x+L) = \phi(x)$.

In addition, the pressure in the fluid at the surface must be atmospheric. The mathematical formulation of these can be found in all the more comprehensive wave texts, e g, Phillips (1966), Le Mehaute (1976), Mei (1989) or Dean and Dalrymple (1991). For the present purposes it suffices to note that the pressure condition at the surface ($p=0$ for $z=\eta$) according to the Bernoulli equation (1.4.2) is

$$\eta = \frac{1}{g}\frac{\partial \phi}{\partial t} \tag{1.4.4}$$

when the second order terms u^2 and w^2 can be neglected.

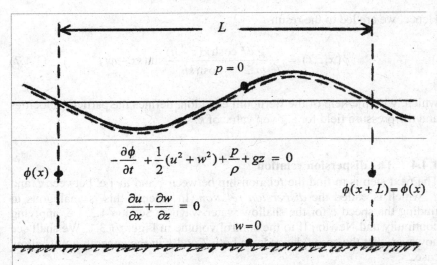

Figure 1.4.1: The field equations are Newton II and the continuity equation. The boundary conditions are $p=0$ @ $z=\eta$, $w=0$ @ $z=-h$ and x-periodicity: $\phi(x+L) = \phi(x)$.

1.4.3 The form of the solution

The method of solving this kind of problems is educated guessing of a potential that will satisfy the field equations and the boundary conditions. We seek a solution with sinusoidal surface variation

$$\eta(x,t) = \frac{H}{2}\cos(kx - \omega t) \qquad (1.3.6)$$

This agrees with the dynamic surface boundary condition (1.4.4) if the potential has the form

$$\phi(x,z,t) = -\frac{g}{\omega}\frac{H}{2}f(z)\sin(kx - \omega t) \qquad (1.4.5)$$

where the function $f(z)$ must equal unity at the surface ($z=0$ for $H \to 0$) and ϕ must satisfy the field equations and the boundary conditions. This is all satisfied with

$$f(z) = \frac{\cosh k(z+h)}{\cosh kh} \qquad (1.4.6)$$

Hence, we are led to the result

$$\phi(x,z,t) = -\frac{g}{\omega}\frac{H}{2}\frac{\cosh k(z+h)}{\cosh kh}\sin(kx - \omega t) \qquad (1.4.7)$$

which, with the help of the Bernoulli equation, defines the particle velocities and the pressure field for a given value of k.

1.4.4 The dispersion relation

The next step is to find the relationship between k and ω, i e, between L and T, which is called the *dispersion relation*. In essence this is analogous to finding the speed c for the shallow water wave in Section 1.3, i e, applying continuity and Newton II to the control volume in Figure 1.3.1. We shall see however, that the speed depends on both T and h in the general, non-shallow case.

From the potential (1.4.7) we can derive an expression for the horizontal water particle velocities:

$$u(x,z,t) = -\frac{\partial \phi}{\partial x} = \frac{gk}{\omega}\frac{H}{2}\frac{\cosh k(z+h)}{\cosh kh}\cos(kx - \omega t) \qquad (1.4.8)$$

We can now apply continuity (Newton II has already been invoked in getting the relation (1.4.4) between η and ϕ from the Bernoulli equation) to the control volume in Figure 1.3.1 but without the shallow water assumptions.

<u>Continuity:</u> $\qquad \dfrac{\partial \eta}{\partial t}\delta_x = \displaystyle\int_{-h}^{\eta}-\dfrac{\partial u}{\partial x}\delta_x dz \qquad (1.4.9)$

The expressions (1.3.6) and (1.4.8) for η and u are used and the upper integration limit η is replaced with 0 (insignificant error when $\eta \ll h$). Noting that

$$\int_{-h}^{0}\frac{\cosh k(z+h)}{\cosh kh}dz = \frac{1}{k}\tanh kh \qquad (1.4.10)$$

we get

$$kh \tanh kh = \frac{\omega^2}{g} h \qquad (1.4.11)$$

which is the dispersion relation for sine waves.

As we shall see in the following sections, all sine wave quantities depend on the water depth through the dimensionless kh and on the elevation through $kz = kh(z/h)$. Hence, convenient methods for finding kh from the dispersion relation are useful.

Since $\tanh x \to 1$ for $x \to \infty$, the deep water (subscript "o") version of the dispersion relation can be reduced to

$$k_o = \frac{\omega^2}{g} = \frac{4\pi^2}{gT^2} \qquad (1.4.12)$$

Hence, the dispersion relation (1.4.11) may alternatively be written as

$$kh \tanh kh = k_o h \qquad (1.4.13)$$

In the shallow water limit, we use the fact that, $\tanh x \to x$ for $x \to 0$ by which, Equation (1.4.13) becomes $(kh)^2 = k_o h$, i e,

$$kh = \sqrt{k_o h} \text{ for } k_o h \to 0 \qquad (1.4.14)$$

For intermediate $k_o h$ a useful approximation of the form

$$kh = \sqrt{k_o h} \, [1 + \alpha k_o h + \beta (k_o h)^2 + ...] \qquad (1.4.15)$$

can be developed. The coefficients α and β are found by inserting the expansion (1.4.15) into (1.4.13) and assuring identity for all powers of $k_o h$. The result is

$$kh = \sqrt{k_o h} [1 + \frac{1}{6} k_o h + \frac{11}{360} (k_o h)^2 + ...] \qquad (1.4.16)$$

This expression is within 1% of the exact solution to (1.4.13) for $k_o h < 2.56$. See Figure 1.4.2. It does however become too inaccurate in deep water. At that end the alternative approximation

Figure 1.4.2:
Relative error of various explicit approximations for *kh*. An error of 1% is generally acceptable in view of the differences between sine waves and real, physical water waves.

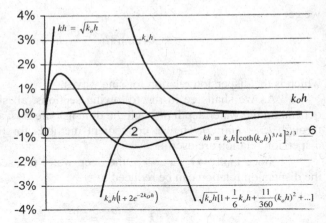

$$kh = k_o h[1 + 2e^{-2k_o h}] \qquad (1.4.17)$$

may be used.

Based on the expansion (1.4.16) and Taylor series for the relevant functions, similar expansions can be obtained for other sine wave related functions. For example, inserting (1.4.16) into $\tanh x = x - \frac{1}{3}x^3 + \frac{2}{15}x^5 + ...$ leads to

$$\tanh kh = \sqrt{k_o h}\,(1 - \frac{1}{6}k_o h + ...) \qquad (1.4.18)$$

This and other similar formulae are listed in the table on page 45 together with their 1% error limits. All of these formulae are in the form of the shallow water or long wave approximation times a Taylor expansion in $k_o h$. In this form they show directly the dependence on water depth and are easy to differentiate or integrate with respect to h.

Coastal engineering applications do not always require all the terms. For example, (1.4.18) may, within 1% accuracy, be reduced to the shallow water expression $\tanh kh \approx \sqrt{k_o h}$ when $\frac{1}{6}k_o h < 0.01$.

30

Most older textbooks and the Shore Protection Manual include so-called wave tables, where common sine wave quantities are listed as functions of $h/L_o = k_o h/2\pi$.

An alternative formula

$$kh = k_o h \left[\coth\left(k_o h\right)^{3/4} \right]^{2/3} \qquad (1.4.19)$$

which conveniently applies over the full range $0 < k_o h < \infty$ with less than 1.5% error has been suggested by Fenton and McKee (1989).

1.4.5 The wave length

In most practical wave calculations the wave period and the water depth will be known while the wave length and hence k need to be calculated. In shallow water the wave length is calculated

$$L_{\text{shallow}} = c_{\text{shallow}}T = \sqrt{gh}\, T \text{ for } k_o h \ll 1 \qquad (1.4.20)$$

In deep water, the wave length is given by

$$L_o = \frac{gT^2}{2\pi} \qquad (1.4.21)$$

This can be derived from the deep water wave number given by (1.4.12) and the general relation $k=2\pi/L$.

Example 1.4.1:

From Equation (1.4.21) we can calculate the wave length in deep water of an 8 second wave. We find $L_{o,T=8s} = 100$m.

☺

To find the wave length at intermediate depths we can either solve the dispersion relation iteratively for kh and then find L from $L=2\pi/k$, or alternatively, we may rewrite (1.4.13) as

$$L = L_o \tanh kh = \frac{gT^2}{2\pi}\tanh kh \qquad (1.4.22)$$

and find tanh*kh* from the explicit approximation (1.4.18), as in the following example.

Example 1.4.2:
Find the wave length of an 8 second wave in 10 metres of water. With $g=9.8m/s^2$ we get $k_o h = 0.6294$.

Iterative solution of $kh \tanh kh = 0.6294$ gives $kh=0.8868$ and since $h=10m$, $k = 0.08868$ leading to $L = 2\pi/k = \underline{70.85m}$. Alternatively, using the explicit approximation (1.4.18) gives $\tanh kh \approx \sqrt{.6294}\,(1 - \dfrac{1}{6}0.6294) = 0.710$, and hence, with $L_o = gT^2/2\pi = 99.82m$, $L = L_o \tanh kh \approx 99.82m \times 0.710 = \underline{70.87m}$. ☺

1.4.6 The wave speed

We saw above that the deep water wavelength is given by $L_o = gT^2/2\pi$. It then follows from the general $L = cT$ that the deep water wave speed is given by

$$c_0 = gT/2\pi \qquad (1.4.23)$$

At intermediate depths the dispersion relation yields, as for the wave length,

$$c = c_o \tanh kh \approx \frac{gT}{2\pi}\sqrt{k_o h}\,(1 - \frac{1}{6}k_o h + ...) = \sqrt{gh}\,(1 - \frac{1}{6}k_o h + ...)$$

$$(1.4.24)$$

In Section 1.3 we found the shallow water wave speed

$$c_{shallow} = \sqrt{gh} \qquad (1.4.25)$$

which is indeed the limiting value of (1.4.24) for $k_o h \to 0$.

If the waves are defined through their length and hence through kh, e g , if L can be measured off a photograph, while the period is perhaps unknown, it is useful to rewrite the dispersion relation as

$$c = \sqrt{\frac{g}{k}\tanh kh} \qquad (1.4.26)$$

The previous derivations have tacitly neglected surface tension, which may be important for very short waves. Its significance compared with gravity can be judged through the more general version of (1.4.26), the dispersion relation for capillary waves:

$$c = \sqrt{\left(\frac{g}{k} + \frac{\sigma}{\rho}k\right)\tanh kh} \qquad (1.4.27)$$

where $\sigma \approx 0.07$N/m is the surface tension of water. For example, for $L=1$cm, the first term in the bracket is 0.016m^2/s^2, while the second term is 0.044 m^2/s^2.

1.4.7 The water particle velocities

The water particle velocities are found from the velocity potential (1.4.7) as done for u in Section 1.4.4 and then using the dispersion relation (1.4.11) to simplify the expressions. The results are

$$u(x,z,t) = \hat{u}\cos(kx - \omega t) = \omega\frac{H}{2}\frac{\cosh k(z+h)}{\sinh kh}\cos(kx - \omega t) \qquad (1.4.28)$$

$$w(x,z,t) = -\hat{w}\sin(kx - \omega t) = -\omega\frac{H}{2}\frac{\sinh k(z+h)}{\sinh kh}\sin(kx - \omega t) \qquad (1.4.29)$$

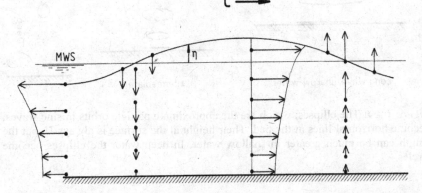

Figure 1.4.3: The wave induced horizontal velocities $u(x,z,t)$ are in phase with the surface elevation $\eta(x,t)$. The vertical velocities $w(x,z,t)$ are in phase with $\partial\eta/\partial t$.

By comparison with (1.3.6) this shows that $u(x,z,t)$ is proportional to $\eta(x,t)$, i e, the maximum horizontal velocity at a point occurs when the wave crest is passing. Similarly we see that the vertical particle velocity is proportional with $\partial\eta/\partial t$. See Figure 1.4.3. The timing relations between η and u,w are different if the waves do not propagate with constant form as shown in Section 4.5 for decaying waves.

We note also that the velocity amplitudes $\hat{u}(z)$ and $\hat{w}(z)$ are both maximum at the surface: $\hat{u}_{\text{surface}} = \omega\dfrac{H}{2}\coth kh$ and $\hat{w}_{\text{surface}} = \omega\dfrac{H}{2}$.

The particle paths corresponding to (1.4.28) and (1.4.29) are not quite closed. However, they do, apart from a second order forward drift (see Section 1.4.14), correspond to elliptical motion with half axes

$$\alpha = \frac{\hat{u}}{\omega} = \frac{H}{2}\frac{\cosh\ k(z+h)}{\sinh\ kh} \tag{1.4.30}$$

$$\beta = \frac{\hat{w}}{\omega} = \frac{H}{2}\frac{\sinh k(z+h)}{\sinh kh} \tag{1.4.31}$$

The shape of these ellipses varies over the depth and between shallow and deep water in the manner shown by Figure 1.4.4.

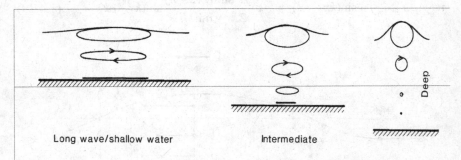

Long wave/shallow water   Intermediate

Figure 1.4.4: The ellipses, which are the approximate particle orbits in sine waves, become horizontal lines at the bed. Their height at the surface is always H, but the length can be much greater in shallow water. In deep water the ellipses become circles.

1.4.8   The pressure field

A leading order expression for the pressure is obtained from the Bernoulli equation (1.4.2) by neglecting the non-linear velocity terms:

$$p(x,z,t) = -\rho gz + \rho \frac{\partial \varphi}{\partial t}$$

$$= -\rho gz + \rho g \frac{H}{2} \frac{\cosh k(z+h)}{\cosh kh} \cos(kx - \omega t)$$

$$= -\rho gz + \hat{p} \cos(kx - \omega t) \tag{1.4.32}$$

According to Equation (1.4.32), the mean pressure is clearly $-\rho gz$. However, this is only true if the point considered is always under water. For points between the trough and the crest, the pressure will be zero (not negative) while the surface is below, see Figure 1.4.5.

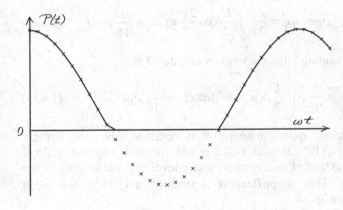

Figure 1.4.5: Pressure variation at a point between the wave trough and the wave crest. The pressure is constant (= atmospheric) while the surface is below.

Correspondingly, the time-mean pressure force per metre of wave crest is greater than $1/2\rho gh^2$, see also Section 1.4.13.

The first part of (1.4.32) is the hydrostatic pressure corresponding to the mean water level. The second part is the dynamic pressure

$$p^+(x,z,t) = \hat{p}\cos(kx - \omega t) = \rho g\eta \frac{\cosh k(z+h)}{\cosh kh} \quad \text{for } z<\eta \tag{1.4.33}$$

35

Example 1.4.3:
Find the amplitude of the pressure fluctuations at the bed under a wave with
$(T,h,H) = $ (8s, 10m, 1.5m).

In Example 1.4.2 we found that kh=0.887 for this (T,h). Hence (1.4.31)
gives the amplitude

$$\hat{p}\big|_{z=-h} \;=\; \rho g \frac{H}{2}\frac{\cosh(0)}{\cosh(0.887)} \;=\; 1025\times 9.8\frac{1.5}{2}\frac{1}{1.420} \;=\; \underline{\underline{5.31 kPa}}$$

The pressure signal at the bed is thus reduced by a factor 1.420 compared to that
near the MWL.   ☺

1.4.9   The energy density
The wave motion represents mechanical energy, which is evenly shared
between kinetic and potential energy. The potential energy corresponding to
the surface elevation η over a length δ_x is $dE_{pot}=1/2\rho g\eta^2\delta_x$, corresponding to
the gravitational energy of the mass/metre $\rho\eta\delta_x$ lifted the height $\eta/2$.
Averaging over the length of a sine wave gives the average potential energy

$$E_{pot} \;=\; \frac{1}{L}\int_0^L \frac{1}{2}\rho g\eta^2\,dx \;=\; \frac{\rho g}{2L}\int_0^L (\frac{H}{2}\cos\frac{2\pi}{L}x)^2 dx \;=\; \frac{1}{16}\rho g H^2 \qquad (1.4.34)$$

The corresponding kinetic energy is calculated as

$$E_{kin} \;=\; \frac{1}{L}\int_0^L\int_{-h}^{\eta} \frac{1}{2}\rho(u^2+w^2)dz\,dx \;\approx\; \frac{1}{16}\rho g H^2 \qquad (1.4.35)$$

The right-hand, approximate result is obtained when the particle
velocity expressions (1.4.28) and (1.4.29) are inserted and the vertical
integration is only extended to the mean water level z=0 rather than to the
actual surface $z=\eta$. This simplification is insignificant under the small
amplitude assumption $\eta \ll h$.

The total energy density (energy per unit area of sea bed) is thus

$$E \;=\; E_{pot}+E_{kin} \;=\; \frac{1}{8}\rho g H^2 \qquad (1.4.36)$$

which, since $\overline{\sin^2\omega t}=\frac{1}{2}$, is the sine wave version of the more general

$$E \;=\; \rho g\overline{\eta^2} \qquad (1.4.37)$$

36

1.4.10 The energy flux

When waves propagate towards a coast and eventually break on the beach they are transporting energy from the generating area towards the beach. The energy flux per metre of wave crest [W/m] through a given vertical section at time t can be calculated as

$$E_f(t) = \int_{-h}^{\eta} [p + \rho g z + \frac{1}{2}\rho(u^2 + w^2)] \, u \, dz \qquad (1.4.38)$$

where the terms in the [ ] are the mechanical energy per unit volume of fluid.

Because of the symmetric nature of u in the sine wave and with the small amplitude approximation $\eta << h$, this boils down to the leading term

$$E_f(t) = \int_{-h}^{0} p^+ u \, dz \qquad (1.4.39)$$

which, with the expressions (1.4.33) and (1.4.28) inserted gives

$$E_f(t) = \frac{1}{8}\rho g H^2 c (1 + \frac{2kh}{\sinh 2kh}) \cos^2(kx - \omega t) \qquad (1.4.40)$$

and, since $\overline{\cos^2} = 1/2$, the time-averaged energy flux per metre of wave crest is

$$E_f = \frac{1}{16}\rho g H^2 c \left(1 + \frac{2kh}{\sinh 2kh}\right) \qquad (1.4.41)$$

This corresponds to a considerable potential source of renewable energy. Consider for example an 11s storm wave with height 5m in deep water. For this wave we get

$$E_f = \frac{1}{16} \times 1025 \times 9.8 \times 5^2 \frac{9.8 \times 11}{2\pi}(1 + 0) = 269 \text{kW/m}$$

which is more power per metre of beach than delivered by most car engines. This figure corresponds to a storm situation and time averaged wave power is generally smaller. However, many locations in the higher latitudes, $>30°$, experience time mean wave power in excess of 60kW/m throughout the year.

1.4.11 Superposition

The theory of small amplitude sine waves is based on the Laplace equation (1.4.3) which is linear and only the linear terms of the Bernoulli equation through (1.4.4). Hence, linear superposition of two sine waves is consistent with the theory of individual sine waves.

Consider thus the combined wave motion

$$\eta(x,t) = \eta_1(x,t) + \eta_2(x,t)$$

$$= \frac{H_1}{2}\cos(k_1 x - \omega_1 t) + \frac{H_2}{2}\cos(k_2 x - \omega_2 t - \varphi)$$

$$= \frac{H_1}{2}\cos k_1(x - c_1 t) + \frac{H_2}{2}\cos k_2(x - c_2 t - \delta_x) \qquad (1.4.42)$$

where φ represents the phase lag of η_2 compared with η_1 at $(x,t) = (0,0)$.

If the waves have the same frequency, i e, $\omega_2 = \omega_1$, and propagate in the same direction so that $k_2 = +k_1$, the result is a new sine wave which propagates with constant form. Its height depends on the phase shift φ, as well as on H_1 and H_2, in accordance with the geometrical rule illustrated in Figure 1.4.6. That is, the height of the combined wave is by the cosine rule

$$H = \sqrt{H_1^2 + H_2^2 + 2H_1 H_2 \cos\varphi} \qquad (1.4.43)$$

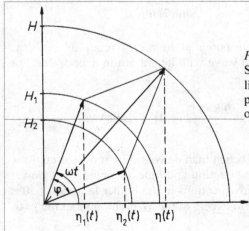

Figure 1.4.6:
Simple harmonic scalar quantities like η_j can be considered as the projections of rotating vectors on one of the coordinate axes.

This representation is useful for finding the total amplitude and phase of a combined wave motion. If the diagram is considered to be in the complex plane, where $H_j e^{i\omega t}$ travels the circle with radius H_j. Recalling Euler's formula: $e^{i\omega t} = \cos \omega t + i \sin \omega t$, we see that $\eta(t) = H_j \cos \omega t = \mathrm{Re}\{H_j e^{i\omega t}\}$.

Combinations of waves of the same frequency $\omega_2 = \omega_1$ but propagating in the opposite direction ($c_2 = -c_1$, $k_2 = -k_1$) will be discussed in the section on wave reflection: Section 1.9.

1.4.12 The group velocity
To develop the concept of group velocity, consider now two waves with slightly different frequencies propagating in the same direction

$$\eta(x,t) = \eta_1(x,t) + \eta_2(x,t) = \frac{H}{2}\cos(k_1 x - \omega_1 t) + \frac{H}{2}\cos(k_2 x - \omega_2 t) \quad (1.4.44)$$

This combination corresponds to a wave pattern like the one shown in Figure 1.4.7.

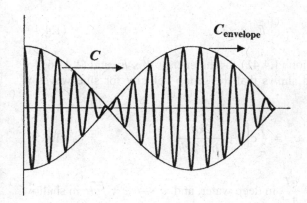

Figure 1.4.7:
The envelope propagates with the velocity c_{envelope}, which is generally smaller than the speed c of the individual waves.

It follows from the trigonometric identity $\cos \alpha + \cos \beta = 2\cos(\frac{\alpha - \beta}{2})\cos(\frac{\alpha + \beta}{2})$ that the sum above can be written

$$\eta(x,t) = H\cos(\frac{k_1 - k_2}{2}x - \frac{\omega_1 - \omega_2}{2}t)\cos(\frac{k_1 + k_2}{2}x - \frac{\omega_1 + \omega_2}{2}t) \quad (1.4.45)$$

The short individual waves in Figure 1.4.7 are described by the last factor, while the first cosine-factor describes the moving envelope. Using the general identity $c = \omega/k$, we see that the envelope propagates with the speed

$$c_{envelope} = \frac{\omega_1 - \omega_2}{k_1 - k_2} \qquad (1.4.46)$$

In the limit of $\omega_2 \rightarrow \omega_1$ and $k_2 \rightarrow k_1$ we thus have

$$c_{envelope} \rightarrow \frac{d\omega}{dk} \qquad (1.4.47)$$

The derivative is usually called the group velocity:

$$c_g = \frac{d\omega}{dk} \qquad (1.4.48)$$

Defining the group velocity as the speed with which wave energy is transported, we may write

$$E_f = E c_g \qquad (1.4.49)$$

Using the expression (1.4.41) for the energy flux E_f, and (1.4.36) for the energy density E then shows that the group velocity for sine waves is given by

$$c_g = \frac{1}{2}c\left(1 + \frac{2kh}{\sinh 2kh}\right) \qquad (1.4.50)$$

We note that $c_g \rightarrow \frac{1}{2}c_o = \frac{gT}{4\pi}$ in deep water, and $c_g \rightarrow c = \sqrt{gh}$ in shallow water.

1.4.13 Radiation stress or wave thrust

Several phenomena such as the upward MWS slope towards the beach in surf zones and the occurrence of longshore currents when waves break at an angle to the beach indicate that waves have an ability to push. However, no quantitative models for these phenomena were available before the independent developments of the theory for radiation stress or wave thrust by Longuet-Higgins and Stewart (1962) and by Lundgren (1963).

The radiation stress S_{xx} is the excess pressure force F_p compared to the still water situation, see Figure 1.4.5, plus the momentum flux F_m per unit length of wave crest

$$S_{xx} = F_p + F_m = \overline{(\int_{-h}^{\eta} p\,dz - \frac{1}{2}\rho g h^2)} + \overline{\int_{-h}^{\eta} \rho u^2 dz} \qquad (1.4.51)$$

which, for a small amplitude sine waves gives the leading terms

$$S_{xx} = F_p + F_m = \frac{1}{16}\rho g H^2\left(1 + \frac{4kh}{\sinh 2kh}\right) \qquad (1.4.52)$$

In the y-direction perpendicular to wave propagation there is no momentum flux but the excess pressure force is the same in all directions. Hence the radiation stress in this direction is

$$S_{yy} = F_p = \frac{1}{16}\rho g H^2 \frac{2kh}{\sinh 2kh} \qquad (1.4.53)$$

1.4.14 Mass transport by sine waves
First, consider the issue of mass transport in Eulerian terms, i e, in terms of the velocity variation at fixed points. The expression (1.4.26) for the horizontal particle velocities at a fixed point under sine waves has the form

$$u(x,z,t) = \hat{u}(z)\cos\omega t \qquad (1.4.54)$$

which indicates $\overline{u(x,z,t)} = 0$ for all points. However, this is only true below the wave trough where there is water at all times. At levels between the trough and the crest the surface goes below during the time interval from say t_d to t_u. The result is a positive net flow rate above the trough, which to the leading order ($\sim H^2$) is

$$Q = \int_{z=-H/2}^{H/2} \int_{t_u(z)}^{t_d(z)} \hat{u}(z)\cos\omega t\,dt\,dz \approx \overline{\eta u(x,0,t)} = \frac{gH^2}{8c} \qquad (1.4.55)$$

If net flow of water is prevented, e g, by the end walls of a wave flume, this flow rate must be eliminated by some seaward flow. The required, vertically uniform seaward velocity, over the range $-h<z<0$, is seen to be $-\dfrac{gH^2}{8ch}$.

Alternatively, the net flow can be worked out in Lagrangian terms, i e, by following individual fluid particles. A fluid particle in a sine wave will drift forward, because the positive velocities along the upper part of its orbit are greater than the negative velocities along the lower part, and because the particle tends to "stay with the wave crest" during its forward motion, see Figure 1.4.8.

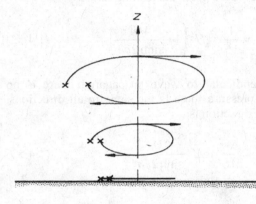

Figure 1.4.8:
The particle orbits in a pure sine wave are not closed.

The resulting, leading order flow rate is again given by (1.4.55). However, the distribution over the depth of the Lagrangian drift is different from that of the Eulerian. The Lagrangian drift is positive throughout, increasing gradually from a finite value at the bed while the Eulerian drift is zero everywhere below the wave trough, see Figure 1.4.9.

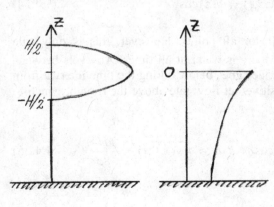

Figure 1.4.9: Eulerian (left) and Lagrangian (right) particle drift velocities corresponding to the sine wave velocities (1.4.28) and (1.4.29). In both cases the corresponding flow rate is $gH^2/8c$.

42

In many cases a beach or the end of the wave flume prohibits net flow of water. What happens then is that, the surface develops a slight slope upwards in the onshore direction, which in turn, drives a steady return flow so that the net flow is zero.

An example of Eulerian mean (×~ averaging over entire wave period, □~ averaging only over the time when there is water) velocities measured in a closed wave flume is shown in Figure 1.4.10.

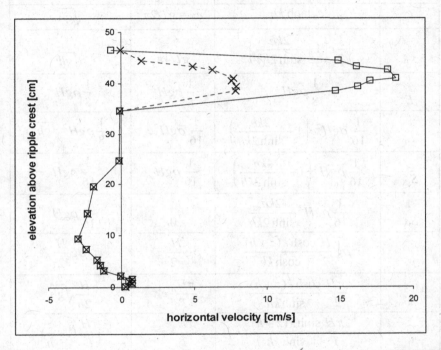

Figure 1.4.10: Eulerian mean velocities in a zero net flow situation. Data from Fredsøe et al. (1999), h= 0.42m, T= 2.5s, H= 13cm, Fixed rippled bed with hydraulic roughness 7.4mm.

Note the shoreward streaming at the bed, which is important for sediment transport, and of different origin to the drift described above. It is a consequence of the horizontal and vertical velocities in the bottom boundary layer not being exactly $\pi/2$ out of phase as in the potential flow. This gives rise to momentum transfer terms of the form $-\overline{\rho u w}$, which drive this "boundary layer streaming" as first explained by Longuet-Higgins (1953, 1956).

43

1.4.15 Summary of sine wave formulae

	General expression	**Deep water**	**Shallow water**
$k_o h$	—	$\dfrac{4\pi^2}{gT^2}h$	—
L	$L_0 \tanh kh$	$L_o = gT^2/2\pi$	$T\sqrt{gh}$
c	$c_0 \tanh kh$	$c_o = gT/2\pi$	$\sqrt{gh}$
c_{g}	$\dfrac{1}{2}c\left(1+\dfrac{2kh}{\sinh 2kh}\right)$	$gT/4\pi$	$\sqrt{gh}$
E	$\dfrac{1}{8}\rho g H^2$	$\dfrac{1}{8}\rho g H_o^2$	$\dfrac{1}{8}\rho g H^2$
E_{f}	$\dfrac{1}{16}\rho g H^2 c\left(1+\dfrac{2kh}{\sinh 2kh}\right)$	$\dfrac{1}{16}\rho g H_o^2 c_o$	$\dfrac{1}{8}\rho g H^2 \sqrt{gh}$
S_{xx}	$\dfrac{1}{16}\rho g H^2\left(1+\dfrac{4kh}{\sinh 2kh}\right)$	$\dfrac{1}{16}\rho g H_o^2$	$\dfrac{3}{16}\rho g H^2$
S_{yy}	$\dfrac{1}{16}\rho g H^2 \dfrac{2kh}{\sinh 2kh}$	0	$\dfrac{1}{16}\rho g H^2$
$\hat{p}^+$	$\rho g \dfrac{H}{2}\dfrac{\cosh k(z+h)}{\cosh kh}$	$\rho g \dfrac{H_o}{2}e^{k_o z}$	$\rho g \dfrac{H}{2}$
$\hat{u}$	$\dfrac{\pi H}{T}\dfrac{\cosh k(z+h)}{\sinh kh}$	$\dfrac{\pi H_o}{T}e^{k_o z}$	$\dfrac{H}{2h}\sqrt{gh}$
$\hat{w}$	$\dfrac{\pi H}{T}\dfrac{\sinh k(z+h)}{\sinh kh}$	$\dfrac{\pi H_o}{T}e^{k_o z}$	$\dfrac{\pi H}{T}(1+\dfrac{z}{h})$
α	$\dfrac{H}{2}\dfrac{\cosh k(z+h)}{\sinh kh}$	$\dfrac{H_o}{2}e^{k_o z}$	$\dfrac{HT}{4\pi}\sqrt{\dfrac{g}{h}}$
β	$\dfrac{H}{2}\dfrac{\sinh k(z+h)}{\sinh kh}$	$\dfrac{H_o}{2}e^{k_o z}$	$\dfrac{H}{2}(1+\dfrac{z}{h})$
K_{s}	$\left[\left(1+\dfrac{2kh}{\sinh 2kh}\right)\tanh kh\right]^{-1/2}$	1	$(4k_o h)^{-1/4}$

Similar formula related to simple sine wave refraction are given in Section 1.7.5.

1.4.16 Explicit approximations to sine wave formulae

Quantity	Explicit approximation	$k_o h$ limit for 1% error
kh	$\sqrt{k_o h}\left(1+\dfrac{1}{6}k_o h+\dfrac{11}{360}(k_o h)^2\right)$	$k_o h < 2.72$
kh	$k_o h\left(1+2e^{-2k_o h}\right)$	$1.75 < k_o h$
c	$\sqrt{gh}\left(1-\dfrac{1}{6}k_o h\right)$	$k_o h < 1.62$
L	$T\sqrt{gh}\left(1-\dfrac{1}{6}k_o h\right)$	$k_o h < 1.62$
c_g	$\sqrt{gh}\left(1-\dfrac{1}{2}k_o h+\dfrac{7}{72}(k_o h)^2\right)$	$k_o h < 2.09$
K_s	$\dfrac{1}{\sqrt[4]{4k_o h}}\left(1+\dfrac{1}{4}k_o h+\dfrac{13}{228}(k_o h)^2\right)$	$k_o h < 1.34$
$\dfrac{1}{\cosh kh}$	$1-\dfrac{1}{2}k_o h+\dfrac{1}{24}(k_o h)^2+0.01(k_o h)^3$	$k_o h < 1.60$
$\dfrac{1}{\sinh kh}$	$\dfrac{1}{\sqrt{k_o h}}\left(1-\dfrac{1}{3}k_o h\right)$	$k_o h < 1.54$
$S_{xx}(H)$	$\dfrac{3}{16}\rho g H^2\left(1-\dfrac{4}{9}k_o h+\dfrac{8}{135}(k_o h)^2\right)$	$k_o h < 0.83$
$S_{xx}(H_o)$	$\dfrac{3}{32}\rho g H_o^2\dfrac{1}{\sqrt{k_o h}}\left(1+\dfrac{1}{18}k_o h\right)$	$k_o h < 0.94$
Setdown (α_o=0)	$-\dfrac{1}{32}k_o H_o^2(k_o h)^{-1.5}\left(1-\dfrac{1}{6}k_o h\right)$	$k_o h < 0.30$
$\dfrac{1}{k_o}\dfrac{1}{c}\dfrac{dc}{dh}=\dfrac{1}{c}\dfrac{dc}{dk_o h}$	$\dfrac{1}{2k_o h}\left(1-\dfrac{1}{3}k_o h-\dfrac{1}{12}(k_o h)^2\right)$	$k_o h < 0.61$

The 1% error limits for the shallow water expressions, which correspond to the leading terms only, can be inferred from the coefficient to $k_o h$ in these formulae. For example, the shallow water expressions for c and L will be within 1% relative error as long as $k_o h/6 < 0.01$, i e, as long as $k_o h < 0.06$.

1.5 WAVES OF FINITE HEIGHT

1.5.1 Introduction

In shallow water, all shapes are possible for small amplitude waves. In arbitrary depths, the solutions to the small amplitude equations are sine waves. As the wave height becomes finite and increases, the shape of the possible steady waves changes. Compared with sine waves, the crest section becomes narrower, particularly in shallow depths. See Figure 1.5.1. The highest waves have sharp crests.

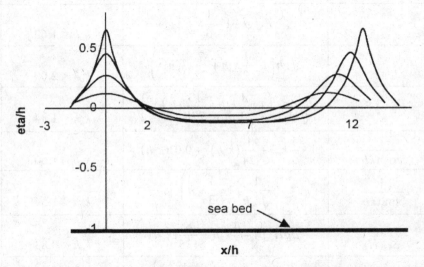

Figure 1.5.1: For given (T,h) the waves get longer and travel faster with increasing wave height. Concurrently, the crests get sharper and the troughs get flatter. Examples from Dean (1974); h/L_o=0.05, H/h = 0.19, 0.39, 0.58, 0.78.

We see that wavelength and celerity increase with wave height: The waves are amplitude dispersive. In shallow water approximately as $c = c_{sine}(1+0.38H/h)$ and in deep water approximately as $c_o = c_{o,sine}[1+11(H_o/L_o)^2]$ compared with sine waves.

The theories for small amplitude shallow water waves and, for sine waves in arbitrary depth, were obtained under the assumptions that the wave height is small in both of the senses $H/h \ll 1$ and $H/L \ll 1$. That is, neither could the waves be too steep nor could the height be a considerable fraction of the depth.

Under the steepness restriction $H/L \ll 1$, the non-linear velocity terms in the Bernoulli equation can be neglected and the dynamic free surface boundary condition

$$-\frac{\partial \phi}{\partial t} + \frac{1}{2}(u^2 + w^2) + g\eta = 0 \qquad (1.5.1)$$

expressing that the pressure at $z=\eta$ must be zero, becomes linear. Similarly, the kinematic free surface boundary condition

$$w = \frac{\partial \eta}{\partial t} + u\frac{\partial \eta}{\partial x} \qquad (1.5.2)$$

becomes linear if the last (non-linear) term is ignored.

Relaxing the steepness criterion ($H/L \ll 1$) then obviously means that solutions based on the linearised free surface boundary conditions become invalid.

When the condition $H/h \ll 1$ is relaxed, the approximation

$$\int_{-h}^{\eta} \circ \, dz \approx \int_{-h}^{o} \circ \, dz \qquad (1.5.3)$$

which has been applied to get the dispersion relation and all integral properties for sine waves breaks down.

1.5.2 The non-linear shallow water equations

Maintaining the shallow water assumptions: that u is uniform in the vertical and p is hydrostatic, the basic principles applied to the control volume in Figure 1.3.1 give

Newton II: $\qquad \rho h \delta_x \dfrac{du}{dt} = -\rho g \delta_\eta h \qquad (1.3.1)$

or with $\dfrac{du}{dt} = \dfrac{\partial u}{\partial t} + u\dfrac{\partial u}{\partial x}$

47

$$\frac{\partial u}{\partial t} + u\frac{\partial u}{\partial x} = -g\frac{\partial \eta}{\partial x} \qquad (1.5.4)$$

Dropping the approximation (1.5.3) adds a term in the continuity equation compared with (1.3.3), i e,

Continuity: $$\frac{\partial \eta}{\partial t} = -\frac{\partial}{\partial x}[u(h+\eta)] \qquad (1.5.5)$$

These are called the nonlinear shallow water equations (NLSWE). No steady solutions, i e, solutions that propagate with constant form exist for these equations. Solutions behave in the way shown in Figure 1.5.2, which is qualitatively similar to the deformation of real waves up to breaking on a slope. A set of actual measurements is shown in Figure 1.7.1, and a detailed review of the applicability of the NLSWEs to coastal engineering is given by Brocchini and Dodd (2008).

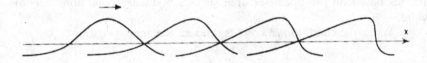

Figure 1.5.2: The nonlinear shallow water equations generate waves that change shape as they go, even on a horizontal bed. After Svendsen and Jonsson (1976).

This deformation corresponds to individual parts of the wave propagating with the "local wave speed" $\sqrt{g(h+\eta)}$ making the crest overtake the slower parts in front and eventually creating a vertical front.

1.5.3 Bores

When the wave front becomes vertical, formulations that contain terms like $\frac{\partial \eta}{\partial x}$ break down. However, shock waves, i e, waves with more or less vertical fronts do exist both in the swash zone, on beaches, and in estuaries. See Figures 2.2.4 and 2.4.1. They are called bores and can be treated as travelling hydraulic jumps, see e g, Lighthill (1978) pp. 181–183 and Liggett (1994) pp. 290–293.

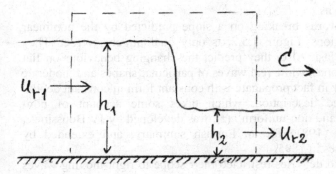

Figure 1.5.3: Control volume following a steady bore. The velocities u_{r1} and u_{r2} are relative to the control volume, i e, relative to the bore front.

The speed of the steady bore in Figure 1.5.3 can be found from the momentum and continuity equations applied to the shown control volume. The momentum equation gives

$$0 = \frac{1}{2}\rho g[h_1^2 - h_2^2] + \rho h_1 u_{r1}^2 - \rho h_2 u_{r2}^2 \qquad (1.5.6)$$

while continuity gives $u_{r1}h_1 = u_{r2}h_2$ and hence

$$u_{r1} = \frac{h_2}{h_1} u_{r2} \qquad (1.5.7)$$

which lead to the general result

$$u_{r2} = \pm\sqrt{\frac{h_1}{h_2} g \frac{h_1 + h_2}{2}} \qquad (1.5.8)$$

For the special case, where the water in front of the bore is at rest relative to the ground, we have $c = -u_{r2}$ and hence get

$$c = \sqrt{\frac{h_1}{h_2} g \frac{h_1 + h_2}{2}} \qquad (1.5.9)$$

That is, bores moving into quiescent water do so with a speed $\sqrt{h_1/h_2}$ times the speed of a small amplitude wave in the average depth.

1.5.4 Cnoidal waves

The deformation towards breaking on a slope predicted by the nonlinear shallow water equations, Figure 1.5.2, is only qualitatively correct. They make it happen too fast. Also, they predict the changing behaviour on flat beds as well as on slopes, while real waves of particular shapes and moderate but finite height, may in fact propagate with constant form in constant depth.

An improved description, which takes some account of non-hydrostatic pressure and non-uniform $u(z)$, was developed by JV Boussinesq in 1872, see Miles (1981) for an English summary, and extended by Korteweg and de Vries in 1895.

Korteweg and de Vries restricted their scope to right-moving waves only and found the equation

$$\frac{1}{\sqrt{gh}}\frac{\partial \eta}{\partial t}+\frac{\partial \eta}{\partial x}+\frac{3}{2h}\eta\frac{\partial \eta}{\partial x}+\frac{1}{6}h^2\frac{\partial^3 \eta}{\partial x^3} = 0 \qquad (1.5.10)$$

which is called the Korteweg-deVries-equation or the KdV Equation for short.

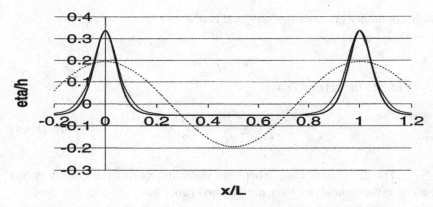

Figure 1.5.4: Wave shape based on 1st order cnoidal theory (light, full line), high order stream function theory, by Dean (1974) (heavy, full line), and sine wave theory (dashed line) for $h/L_o= 0.01$, $H/h= 0.39$.

The first two terms in (1.5.10) correspond to right-moving shallow water waves $\eta(x,t)=f(x-\sqrt{ght})$, while the third term is the leading order term due to finite η/h and the fourth term is the leading order term due to non-hydrostatic pressure $\sim$ finite kh. The KdV Equation has solutions that propagate with constant form, which are called cnoidal waves. Their shape is indicated by Figure 1.5.4.

The existence of these solutions is possible because the 3rd and 4th terms in (1.5.10) can balance each other. The 3rd term, which is shared with the non-linear shallow water equations tends to make "tall parts the waves" move faster while the 4th term make "strongly upward convex parts of the waves" move slower.

Cnoidal wave theory is only valid for long waves, i e, for $h \ll L$. However, it describes the variation of wave height up to breaking much better than sine wave theory, see Figure 1.5.5.

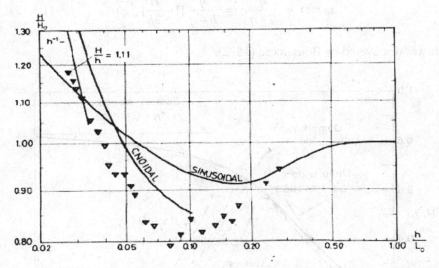

Figure 1.5.5: First order cnoidal theory is better than sine wave theory for predicting wave height variation near breaking on a beach. After Svendsen and Jonsson (1976).

This improved description of shoaling, compared with sine waves, is related to the fact that the narrow crested cnoidal shape, Figure 1.5.4, is more appropriate than a sine curve for waves, which approach breaking cf Figure 1.7.1.

Tables and guidelines with which shoaling of first order cnoidal waves may be calculated can be found in Svendsen and Jonsson (1976) and methods for calculating cnoidal waves up 5th order are described by Fenton (1991).

As indicated by Figure 1.5.6, the limiting form of cnoidal waves for $L/h \to \infty$ is the solitary wave, for which the surface shape is given by:

$$\eta(x,t) = \frac{H}{\cosh^2\left[\sqrt{\frac{3H}{4h^3}}(x-[1+\frac{H}{2h}]\sqrt{gh}\,t)\right]} \qquad (1.5.11)$$

with corresponding depth-uniform fluid velocity

$$u(x,t) = \frac{\eta}{h+\eta}c = \frac{\eta}{h+\eta}[1+\frac{H}{2h}]\sqrt{gh} \qquad (1.5.12)$$

It was discovered by Boussinesq (1872).

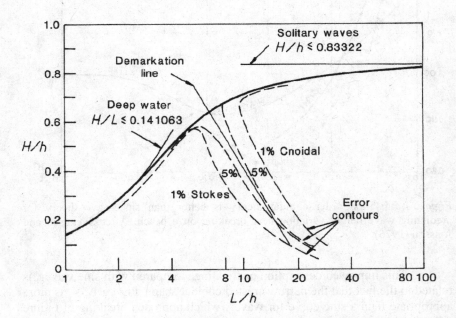

Figure 1.5.6: Outline of the parameter space in which steady wave solutions exist and of the ranges of application for the various theories. Solitary waves are cnoidal waves with $L/h \to \infty$. After Fenton (1990).

1.5.5 Stokes waves

A different type of finite amplitude wave theory was developed by GG Stokes in (1847). He made use of the fact that the sine wave potential 1.4.7 satisfies both continuity in the interior and the bottom boundary condition exactly. This together with the fact that the continuity equation is linear means that a potential made up as a sum of sine wave potentials will automatically satisfy continuity and the bottom boundary condition. The remaining problem is to satisfy the two surface boundary conditions.

The Stokes' scheme is, at second order, to improve the agreement with the surface boundary conditions by adding a potential ϕ_2 corresponding to half the wavelength and half the period. Third order Stokes theory includes another harmonic ϕ_3 with length $L/3$ and period $T/3$ etc. Fenton (1991) gives the coefficients needed for calculation of Stokes waves up to 5th order.

While cnoidal wave theory is only valid for long waves, Stokes theory is only valid for fairly short waves, see Figure 1.5.6.

Fenton (1990) explains how the Stokes approximations become divergent in shallower depths. His demarcation line between cnoidal theory and Stokes theory, which is shown in Figure 1.5.6 is given by

$$L/h = 21.5 \exp[-1.87H/h] \qquad (1.5.13)$$

1.5.6 Fourier expansion methods

The Fourier expansion methods are similar to Stokes waves in that the wave shape is made up from harmonics of lengths $L, L/2, L/3, \ldots, L/N$.

However, while Stokes theory is an analytical successive approximations technique, where the coefficients for successive harmonics are found progressively, the coefficients in the Fourier expansion methods are all determined together by a least squares optimisation of the free surface boundary conditions. With sufficiently many harmonics these methods can be made to give reasonable accuracy through the whole h/L range. The thick solid curve in Figure 1.5.6 is the line below which Fourier expansion methods for steady, irrotational waves have been found to converge.

Fenton (1988) provides a computer code for calculating Fourier expansion waves of high order, and comprehensive results obtained with a 17th order theory are tabulated by Dean (1974).

1.6   NATURAL, IRREGULAR WAVES

1.6.1   Introduction

The waves that are generated in the natural environment are irregular. They have different lengths and different height. Large waves usually come in groups, known to surfers as sets. A sequence of deep water waves measured by a wave rider buoy is shown in Figure 1.6.1.

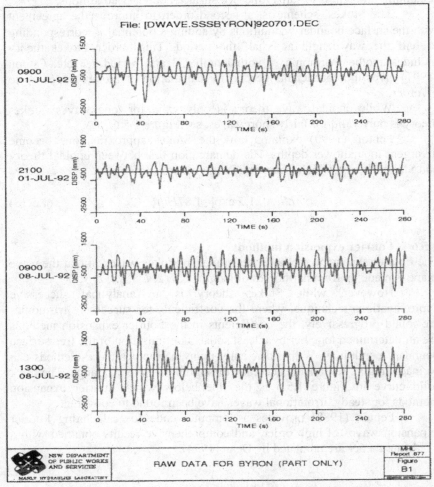

Figure 1.6.1: Time series of water surface elevations recorded by a wave rider buoy off cape Byron. Manly Hydraulics Laboratory Ann Rep 1996–97.

These are surface elevation time series which may be generated by one or more wave trains, e g, swell from the NE and locally generated wind waves from the SE. There is a large literature on natural, irregular waves and a recent comprehensive treatment is given by Holthuijsen (2007).

The first task is to define what we will consider a wave in such a record. Then we can calculate statistics of the waves within that record. We get a set of short term statistics for example the mean period $\overline{T}$, the mean height $\overline{H}$ and the maximum height H_{max} for the record.

These short term statistics will then vary between records taken on different days and long term statistics such as $\overline{H_{max}}$, i e, the average of maximum wave heights or $(H_{max})_{1\%}$, i e, the H_{max} value that has only been exceeded by 1% of all the records are of interest for design purposes.

1.6.2   Definition of individual waves and simple statistics

In a record of irregular waves there are two possible definitions of the wave heights, which correspond to individual waves being bounded by subsequent surface upcrossings or by subsequent downcrossings, see Figure 1.6.2:

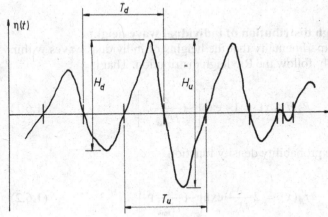

Figure 1.6.2: Either zero upcrossings or zero downcrossings can be used to separate individual waves.

(1) Measure from a trough to the following crest. The result is the *zero downcrossing height*, H_d.

(2) Measure from a crest to the following trough. The result is the *zero upcrossing height*, H_u.

Correspondingly the record represents two sets of wave periods, T_d and T_u. The averages are invariant with respect to the choice of upcrossing versus downcrossing: $\overline{H_u} = \overline{H_d}$ and $\overline{T_u} = \overline{T_d}$. However, that is not the case for other statistics such as the maximum wave height in a record H_{max}, the root mean square wave height $H_{rms} = \sqrt{\overline{H^2}}$ or the significant wave height H_{sig} which is defined as the average height of the largest 1/3 of the waves.

The average zero crossing period for a record is often referred to as $T_z (= \overline{T_u} = \overline{T_d})$.

The surface elevation time series in Figure 1.6.1 are symmetrical about zero and the distribution is well approximated by the normal distribution, $\eta(t)$ is then Gaussian (a Gaussian stochastic process). This is generally the case for waves in deep water. In shallow water where the crests are sharper and narrower and the troughs longer and flatter (as for the waves in Figure 1.5.1) the process is not Gaussian.

1.6.3 The Rayleigh distribution of individual wave heights

It has been found experimentally that the heights of individual waves within a given record usually follow the Rayleigh distribution. That is

$$P\{H \le x\} = 1 - \exp\left[-(\frac{x}{H_{rms}})^2\right] \qquad (1.6.1)$$

corresponding to the probability density function

$$f_R(x) = 2\frac{x}{H_{rms}^2}\exp\left[-(\frac{x}{H_{rms}})^2\right] \qquad (1.6.2)$$

Example 1.6.1:
Find the probability that the next wave is going to have a height of at least $2H_{rms}$.

Assuming the Rayleigh distribution (1.6.1), it follows that $P\{H \geq x\} =$ $\exp[-(\dfrac{x}{H_{\mathrm{rms}}})^2]$ in general, and if we want a wave with twice the rms height we get

$$P\{H \geq 2H_{\mathrm{rms}}\} \;=\; \exp[-(\dfrac{2H_{\mathrm{rms}}}{H_{\mathrm{rms}}})^2] \;=\; e^{-4} \;=\; \underline{\mathbf{0.0183}} \qquad ☺$$

Under the assumption of the Rayleigh distribution, it is possible to estimate special statistics like $H_{1\%}$ by ranking the larger waves and plotting them on a Rayleigh plot. That is a graph where the Rayleigh distribution plots as a straight line. It is devised as follows.

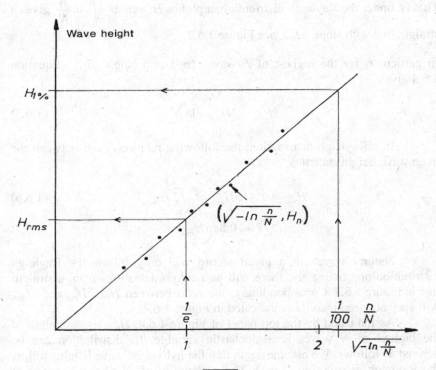

Figure 1.6.3: Plotting H_n versus $\sqrt{-\ln\dfrac{n}{N}}$ for waves in a Rayleigh distribution gives a straight line with slope H_{rms}. Extreme wave heights like $H_{1\%}$ can be derived as shown.

Consider a wave record consisting of N waves, which we have ranked in descending order so that the largest is H_1, the second largest is H_2 and the nth largest H_n etc. The distribution function (1.6.1) then implies

$$P\{H \geq H_n\} = \frac{n}{N} = \exp[-(\frac{H_n}{H_{rms}})^2] \qquad (1.6.3)$$

or by solving for H_n:

$$H_n = H_{rms} \sqrt{-\ln \frac{n}{N}} \qquad (1.6.4)$$

That is, under the Rayleigh distribution, a plot of H_n versus $\sqrt{-\ln \frac{n}{N}}$ gives a straight line with slope H_{rms}, see Figure 1.6.3.

In particular, for the highest of N waves ($n=1$ and height H_{max}) Equation 1.6.4 gives

$$H_{max} = H_{rms} \sqrt{\ln N} \qquad (1.6.5)$$

In a Rayleigh distribution, the following relations exist between the main wave height statistics:

$$H_{sig} = 1.416 H_{rms} \approx \sqrt{2} \, H_{rms} \qquad (1.6.6)$$

$$\bar{H} = 0.886 H_{rms} \qquad (1.6.7)$$

Natural waves in a given record will only follow the Rayleigh distribution approximately, there will be random deviations from a straight line in Figure 1.6.3. Correspondingly, the ratio between H_{rms}, H_{sig} and H_{max} will vary between records as indicated in Figure 1.6.4.

As indicated by the top panel of Figure 1.6.4, H_{max} or the height of the highest of N waves is a stochastic variable. Its distribution can be derived as follows. We assume again that the individual wave heights follow the Rayleigh distribution (1.6.1). The probability of all of the N waves in the record being smaller than a given height x is then

$$P\{H_{max} \leq x\} = P\{\text{All of } N \text{ waves} \leq x\} = (1 - \exp[-(\frac{x}{H_{rms}})^2])^N \qquad (1.6.8)$$

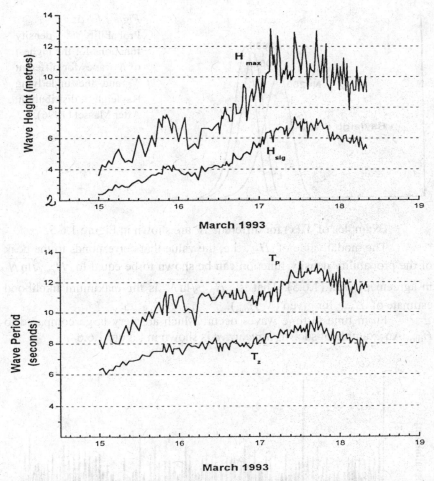

Figure 1.6.4: Height and period statistics from a series of 26.6minute records recorded in 80m of water off Pt Lookout near Brisbane during Cyclone Roger, March 15–19, 1993.

This is the distribution function for the height of the highest of N waves, and the corresponding probability density function is found by differentiation

$$f_N(x) = N(1 - \exp[-(\frac{x}{H_{rms}})^2])^{N-1} 2 \frac{x}{H_{rms}} \exp[-(\frac{x}{H_{rms}})^2] \qquad (1.6.9)$$

59

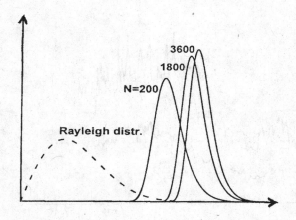

Figure 1.6.5: Probability density functions for the highest of N waves for different N, and the underlying Rayleigh distribution. After Massel (1996).

Examples of $f_N(x)$ for different N are shown in Figure 1.6.5.

The modal value of H_{max}, i e, the value that corresponds to the peak of the probability density function can be shown to be equal to $H_{rms}\sqrt{\ln N}$ in agreement with (1.6.5). That is, $H_{rms}\sqrt{\ln N}$ is the maximum likelihood estimate of H_{max} for given (N, H_{rms}).

From time to time waves occur, which are very large compared to H_{rms}. An example of such a *rogue wave* is shown in Figure 1.6.6.

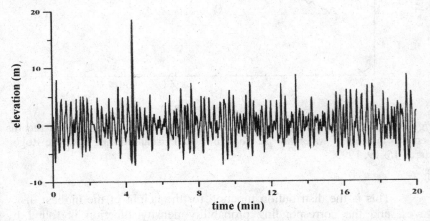

Figure 1.6.6: A 26m rogue or freak wave measured in 70m of water in the North Sea, New Year's Day 1995. While this wave was very large, the most extreme wave in relative terms, had a height of $3H_{sig}$ (6m in a record with $H_{sig}=2$m). After Kharif and Pelinovsky (2003).

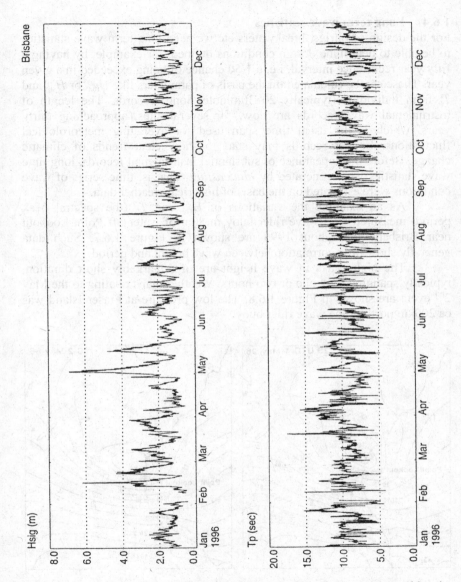

Figure 1.6.7: H_{sig} and spectral peak period measured by a wave rider in approximately 80m of water off Point Lookout, near Brisbane. Note of how short duration the extreme storm events typically are. Data courtesy of the Queensland Environmental Protection Authority.

1.6.4   Long term wave statistics

For the design of oilrigs, breakwaters etc we need long term wave statistics to be able to determine design conditions defined, for example, by having a fifty year recurrence interval, i e, a 1/50 chance of being exceeded in a given year. These can be calculated on the basis of parameters like H_{max} or H_{sig} and T_p from individual (typically 20–30minutes long) records. The length of instrumental wave records are now, for several sites, approaching thirty years, which is the usual time span used for averaging meteorological fluctuations. Longer records may start to show clear trends of climatic change. Before the appearance of substantial instrumental records, long time wave statistics were generated by *hindcasting*. That is, time series of wave conditions were estimated on the basis of historical weather data.

As an example the variations of H_{sig} and T_p (the spectral peak period) measured by a wave rider buoy in 80m of water off Point Lookout near Brisbane throughout 1996 are shown in Figure 1.6.7. Such data generally show little correlation between wave height and period.

The high peaks in wave height are of remarkably short duration, typically spanning only one or two hours. Weather maps relating to the May 2nd event are shown in Figure 1.6.8. The low pressure at Fraser Island was ca 200km north of the wave rider buoy.

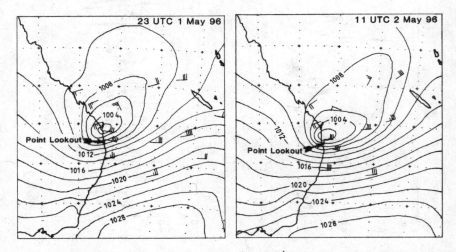

Figure 1.6.8: The weather maps for May 1st and 2nd 1996 when H_{sig} reached 7m at Point Lookout. The position of the wave rider is indicated by the arrow. The stems of the wind markers show direction and the "feathers" give the speed in Knots, 10kn for one full length "feather". Map produced by The Australian Bureau of Meteorology.

For the purpose of summarising long term statistics and possibly enabling extrapolation, exceedence plots like Figure 1.6.9 are often used. When applying exceedance plots it should be checked whether all major events are actually included. That is, that no equipment failure occurred during events that may in fact have been among the most severe.

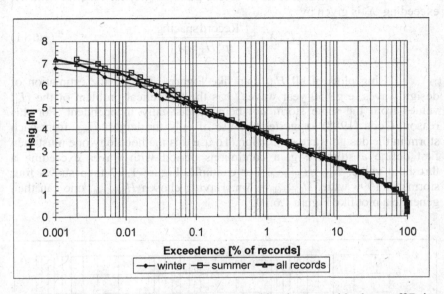

Figure 1.6.9: Exceedence plots for H_{sig} measured by the wave rider buoy off Point Lookout near Brisbane from the period March 1967 to March 2003. Initially the record spacing was 6 hours. More recently, including the handful of most extreme events, the spacing has been 1 hour. Data courtesy of Queensland Environmental Protection Authority.

Hogben (1990) recommends the use of Weibull distributions for extrapolation to extreme conditions. Weibull distributions are the three-parameter family of distributions with the form

$$p\{H_s > X\} = e^{-\left(\frac{X - H_{s0}}{H_{s1} - H_{s0}}\right)^{\gamma}} \qquad (1.6.10)$$

for the exceedence probability. The parameters H_{s0}, H_{s1} and γ may be chosen to obtain the best basis for extrapolation of a given data set. In praxis, the fitting is then done on a plot of $(-\ln p)^{1/\gamma}$ versus H_s in analogy with the Rayleigh plots in Figure 1.6.3. However, even with such a wide family of distributions, the job of extrapolation is not easy, see the example in Figure 1.6.9. The data show no consistent linear trend for $H_{sig} > 6m$, and the choice of a different p makes little difference.

The recurrence interval or return period $T_R(X)$ for records with H_s exceeding X is given by

$$T_R(X) = \frac{\text{Record spacing}}{p\{H_s > X\}} \qquad (1.6.11)$$

Using plots of all H_{sig} data like Figure 1.6.9 for the estimation of design waves, e g "50 year waves" has the theoretical problem of the H_{sig} values from the same storm not being statistically independent. For this reason, Goda (1988) suggested the use of only the largest H_{sig} from each storm in the design wave estimation. To use this methodology one must then first define a storm, e g, a continuous period with waves exceeding a threshold value: $H_{sig} > H_{threshold}$. For example, Figure 1.6.7 then shows four storms in 1996 with $H_{threshold} = 4m$. Having chosen $H_{threshold}$, one can then generate a plot like Figure 1.6.10.

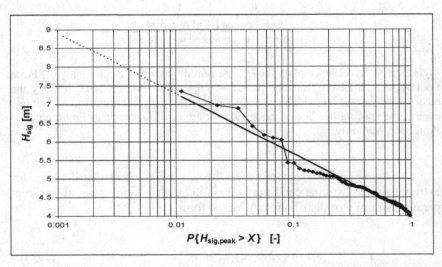

Figure 1.6.10: Peak significant wave heights at Point Lookout, off Brisbane. The dataset contains 88 storms with $H_{sig} > 4m$ spanning the period 30/10 1976 to 30/6 2005. The average spacing of $4m^+$ storms is thus 28.67 years/88 storms = 0.326 years. The straight line corresponds to the best fit Weibull distribution.

Note that also for an extreme value plot like Figure 1.6.10, the best fit straight line is not easy to agree upon. It would clearly be quite different if $H_{threshold}$ were chosen as 5m or 6m instead of 4m, even if the semi-logarithmic format were maintained.

Example 1.6.2:
The data point with $H_{sig} \approx 7m$ in Figure 1.6.10 indicates that, in a given storm 7m is exceeded with a probability of 0.022. In other words, it takes on the average $1/0.022 = 45.5$ storms to exceed $H_s = 7m$.

With an average storm spacing of 0.326 years that means a return period of $T_R(H_{sig}=7m) = 45.5 \times 0.326 = 14.8$ years. Or rather: H_{sig} has the probability $1/14.8 = 0.068$ of getting exceeded in a given year.

If, instead of the data points themselves, one used the straight line corresponding to the best fit Weibull distribution, the estimate would be ca 50% longer, i e, more like 20 years.

Secondly estimate the peak H_{sig} for a 50 year storm? With a storm spacing of 0.326 years this corresponds to $50/.326=153$ storms or a probability of 0.0065. From the straight, dotted line this corresponds to $H_{sig}= 7.7m$. I e, the fifty year storm is, on the basis of the shown Weibull extrapolation, expected to have a peak H_{sig} of 7.7m.

Storm severity is not necessarily just a matter of peak wave height. A more detailed definition/measure of a storm severity, including storm duration and perhaps wave period may be also worthwhile. For example, if the investigation is about wave inundation, the runup scale $\sqrt{HL_o} \propto T\sqrt{H}$, see Section 2.4.3, is of interest and one should then look at the long term statistics of $T_p\sqrt{H_{sig}}$.

1.6.5 Wave spectra
One way of condensing the information of a wave record is in the form of the wave spectrum. That is the distribution of the wave energy among frequencies. The spectrum is evaluated on the basis of the discrete Fourier representation

$$\eta(t) = \sum_1^N R_n \cos(n\omega_{min}t - \phi_n) \qquad (1.6.12)$$

of the surface elevation record.

The minimum angular frequency is

$$\omega_{min} = 2\pi f_{min} = 2\pi \frac{1}{\text{record length}} \qquad (1.6.13)$$

and the maximum angular frequency which can be resolved with a given sampling interval δ_t is 2π times the so-called Nyquist frequency

$$\omega_{max} = N\omega_{min} = 2\pi f_{Nyquist} = 2\pi \frac{1}{2\delta_t} \qquad (1.6.14)$$

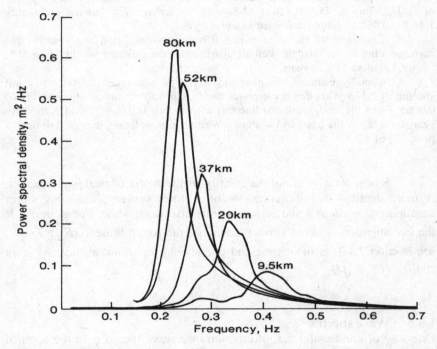

Figure 1.6.11: Spectral shapes measured after different fetch lengths in deep water during the JONSWAP experiment, Hasselmann et al. (1973).

The spectral density function, or simply, the spectrum corresponding to the Fourier series (1.6.12) is defined as

$$S_{\eta\eta}(n\omega_{min}) = \frac{1}{2} R_n^2 \frac{1}{\omega_{min}} \qquad (1.6.15)$$

Various mathematical forms of the spectrum function have been suggested. The most detailed one is the so-called JONSWAP spectrum of Hasselmann et al. (1973)

$$S_{\eta\eta}(\omega) = C_1 g^2 \omega^{-5} \exp\left[-\frac{5}{4}\left(\frac{\omega}{\omega_{peak}}\right)\right] C_2^{C_3} \qquad (1.6.16)$$

where the constants C_1, C_2, and C_3 depend on the wind speed, and the effective fetch over which the waves have been growing. For details, see Hasselman et al. (1973) or Massel (1996). Examples of spectral shapes after different fetch lengths are shown in Figure 1.6.11.

We see that as the waves grow the peak period becomes longer and the spectrum becomes narrower: the waves become increasingly narrow banded.

Most of the work on spectra has focused on deep water and thus avoided the further complications due to wave interactions with the bed. However, some details of the limited depth effects on wave growth can be found in Young and Verhagen (1996).

As opposed to the single peaked spectrum of growing wind waves, the spectrum of shoaling swell waves have multiple peaks corresponding to the higher harmonics of the cnoidal wave shape, cf Figure 1.5.3.

The time averaged energy density $\rho g \overline{\eta^2}$ (cf Equation (1.4.34)) for the wave record is equal to ρg times the area under the spectrum. This follows from Parceval's theorem for a Fourier series like (1.6.12).

$$\overline{\eta^2} = \frac{1}{2}\sum_{n=1}^{N} R_n^2 \qquad (1.6.17)$$

which through the definition (1.6.15) of the spectrum function corresponds to

$$\overline{\eta^2} = \int S_{\eta\eta}(\omega)d\omega \qquad (1.6.18)$$

The nth *spectral moment* is defined as

$$M_n = \int \omega^n S_{\eta\eta}(\omega)d\omega \qquad (1.6.19)$$

Hence, M_o is simply the area under the spectrum. Instead of the peak frequency another frequency defined by the spectrum is sometimes used namely $\omega_{20} = \sqrt{M_2 / M_o}$ corresponding to the priod

$$T_{02} = 2\pi\sqrt{M_0 / M_2} \qquad (1.6.20)$$

The discussion above deals with spectra that are relevant to measurements of $\eta(t)$ at a point for example by a surface piercing staff or a wave rider buoy, which contain no information about the direction of wave propagation. Devices are however becoming available which record directional information as well. Correspondingly a directional spectrum can be calculated in the form

$$S_{\eta\eta}(\omega, \theta) = S_{\eta\eta}(\omega) D(\theta) \qquad (1.6.21)$$

The spreading function usually has the form $D(\theta) = $ constant $\times$ $\cos^{2P}(\theta - \theta_o)$, see for example Massel (1996) or Holthuijsen (2007) for further details.

1.6.6 Generation of waves by wind

The process of wave generation by wind is so complicated that the most fruitful approach to modelling it so far has been empirically via dimensional analysis. The primary parameters that have been sought are H_{sig} and a peak period T_p. The main independent variables are: a wind speed, usually the average speed U_{10} measured 10m above the ocean, the time T_{wind} that this wind speed has prevailed and the fetch F_{wind} over which the waves have been subjected to this wind. Also, the densities of the air ρ_{air} and of the water ρ_w, the surface tension γ, and g must be included. The above parameters suffice for deep water but in shallower depths the water depth h must also be included. The general formulation (neglecting variability of ρ_{air}, ρ_w and γ) in terms of dimensionless variables is for the dimensionless wave height

$$\frac{gH_{sig}}{U_{10}^2} = \Phi_H\left(\frac{gT_{wind}}{U_{10}}, \frac{gF_{wind}}{U_{10}^2}, \frac{gh}{U_{10}^2}\right) \qquad (1.6.22)$$

and for the peak period

$$\frac{gT_p}{U_{10}} = \Phi_T\left(\frac{gT_{wind}}{U_{10}}, \frac{gF_{wind}}{U_{10}^2}, \frac{gh}{U_{10}^2}\right) \qquad (1.6.23)$$

where one or more of the independent variables can be ignored if they become very large. The Shore Protection Manual (US Army Corps of Engineers 1984) and the Coastal Engineering Manual http://chl.erdc.usace.army.mil/chl.aspx?p=s&a=ARTICLES;104 provide very detailed formulae, charts and guidelines for wind wave estimations, including incorporation of temperature and topographical effects into a "wind stress factor" U_A which may be used in the place of U_{10} in order to account for temperature and special topographical effects.

The fact that weather systems are usually not stationary leads to special considerations as illustrated by the following case: On January 6, 2004 the small Pacific Island nation of Niue lost its supermarket and many other buildings on the top of 20m cliffs. The waves, which destroyed the buildings, were generated by Tropical Cyclone Heta, which was estimated to have 30m/s winds at the time, over a fetch of 225km. Based on this, the wind wave charts, gives an H_{sig} of 7 to 8 metres. Not likely to destroy buildings on 20m cliffs.

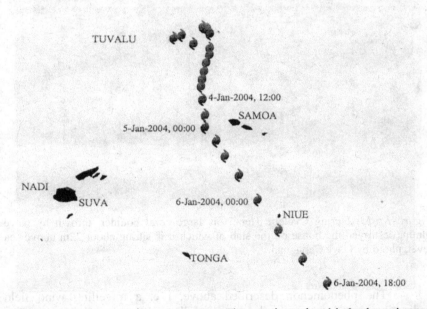

Figure 1.6.12: The track of Cyclone Heta. The way it sped and indeed accelerated towards Niue resulted in the waves that hit Niue being at least twice as big as the waves expected from a similar but stationary wind system. After Callaghan et al. 2007.

However, because the cyclone had been travelling in the same direction as the waves that eventually hit Niue, and with a speed (~20m/s) which approaches the group velocity of the waves, these waves had been exposed to 30m/s winds for about 29 hours. For such waves the wave chart gives H_{sig} of the order 16m, which makes sense in relation to the observed destruction of buildings atop 20m cliffs. Callaghan et al. (2007) give a complete description. Figure 1.6.13 shows the writer next a coral boulder which was, by the same waves, placed on the landward side of the coast road which runs 20–24m above sea level on this stretch.

Figure 1.6.13: During Cyclone Heta, this large coral boulder, thrown by wave runup, destroyed the house on the slab of which it is sitting about 22m above sea level, photo by Dave Callaghan.

The phenomenon described above, i e, a travelling wind field generating very large waves because it follows the waves, who therefore experience a virtual fetch much larger than the diameter of the wind system, is called *fetch enhancement*. It was originally discovered in relation to small (cyclone size) arctic low pressure systems and discussed by Dysthe and Harbitz (1987).

1.7   WAVES CHANGING DUE TO TOPOGRAPHY

1.7.1   Introduction

As waves propagate into shallower water they change both their height and their shape. The height can increase many fold particularly for long swell waves and the crest section gets narrow, cf Figures 1.5.4 and 1.7.1. Waves on a slope deviate from the steady, (horizontal bed) solutions in Figure 1.5.4 by developing a forward leaning shape associated with acceleration skewness: $(du/dt)^3 > 0$, see Figure 1.7.1. This acceleration skewness is important for the associated sediment transport.

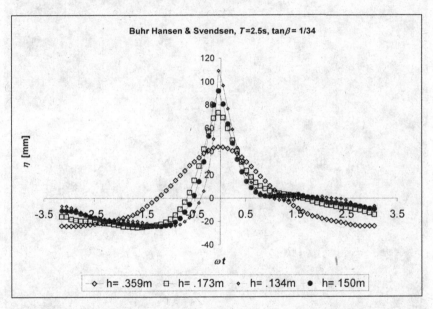

Figure 1.7.1: Time series of surface elevation in different depths for waves shoaling on a slope. Data from Buhr Hansen and Svendsen (1979).

Strictly speaking, the modelling of this wave transformation requires an unsteady wave theory since the waves obviously do not propagate with constant form. However, no closed form solutions exist for unsteady waves. They can only be modelled numerically. The theory for waves in such variable depth is comprehensively treated by Dingemans (1995). The practical approach is usually to model the shoaling process as a sequence of steady waves in equilibrium with discrete local depths.

If the waves propagate at an angle with bottom contours, their direction will change like light entering a glass prism at an angle, see Figure 1.7.2. The process is analogous to the refraction of light, i e, the direction change is due to the change of wave speed with depth.

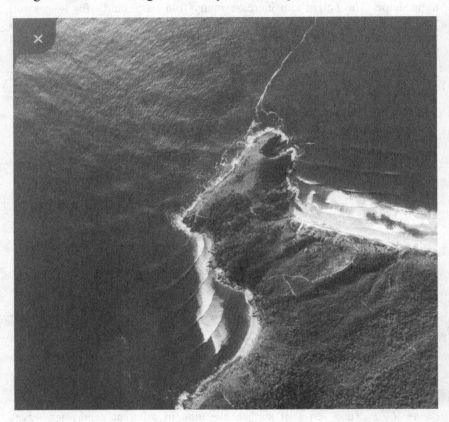

Figure 1.7.2: Refraction: Swell waves manage to break with crests almost parallel to both beaches. The sharp curvature of the wave crests near the bottom left end of the headland is caused by diffraction. Photo supplied by The State of Queensland, Department of Natural Resources and Water [1972].

1.7.2   Shoaling of sine waves

Shoaling is the process of wave height change due to depth decrease. It is modelled on the basis of energy flux considerations as illustrated in Figure 1.7.3.

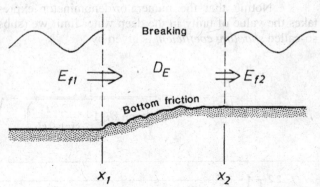

Figure 1.7.3: As the waves propagate through the control volume, the energy flux can only be changed by the energy dissipation D_E [W/m^2]. The dissipation can be due to breaking, opposing wind stress and bottom friction.

In general we have

$$\frac{dE_f}{dx} = \frac{d}{dx}(E c_g) = -D_E \qquad (1.7.1)$$

However, if the energy dissipation D_E [W/m^2] is negligible, this consideration yields

$$E_f = E c_g = \text{constant} \qquad (1.7.2)$$

which with the formulae for energy density and group velocity from sine wave theory becomes

$$\left[\frac{1}{8}\rho g H^2\right]\left[\frac{1}{2}c_o \tanh kh\left(1+\frac{2kh}{\sinh 2kh}\right)\right] = \text{constant} \qquad (1.7.3)$$

For the wave heights at the two depths h_1 and h_2 we get

$$\frac{H_2^2}{H_1^2} = \frac{\tanh k_1 h_1\left(1+\dfrac{2k_1 h_1}{\sinh 2k_1 h_1}\right)}{\tanh k_2 h_2\left(1+\dfrac{2k_2 h_2}{\sinh 2k_2 h_2}\right)} \qquad (1.7.4)$$

73

Noting that the numerator/denominator expresssion on the right takes the value of unity in the deep water limit we (subscript o) see that the so-called *shoaling coefficient* is given by

$$\frac{H}{H_o} = K_s = \frac{1}{\sqrt{\tanh kh\left(1 + \dfrac{2kh}{\sinh 2kh}\right)}} = \sqrt{\frac{\frac{1}{2}c_o}{c_g}} \qquad (1.7.5)$$

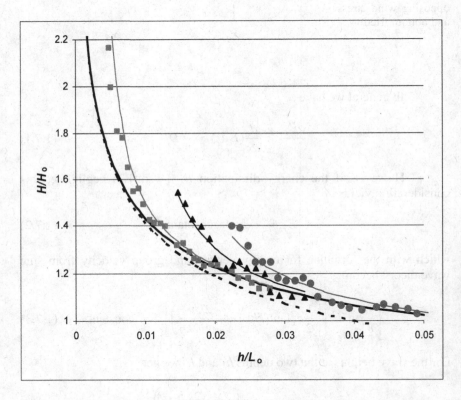

Figure 1.7.4: Wave height variation for waves shoaling on a slope. The thin lines are Equation (1.7.7) compared with data from Sato et al. (1992): ■ H_o/L_o=0.001, ▲ H_o/L_o=0.005, ● H_o/L_o=0.01.The thick line is Equation (1.7.6) i e, sine wave theory corresponding to H_o/L_o→0. The dotted line is its shallow water approximation $\dfrac{H}{H_o} = \dfrac{1}{\sqrt[4]{4k_o h}}$.

This has the convenient explicit approximation

$$\frac{H}{H_o} = K_s = \frac{1}{\sqrt[4]{4k_oh}}\left(1+\frac{1}{4}k_oh+\frac{13}{228}(k_oh)^2\right) \qquad (1.7.6)$$

which is within 1% of (1.7.5) for $k_oh<1.34$.

    The latter expression shows that for sine waves, which are not being reflected, the wave height increases as $h^{-1/4}$ in shallow water. However, as shown by the data in Figure 1.7.4, this small amplitude result is an underestimate for waves near breaking for which H/h is always of the order unity. The trend near breaking on a beach is closer to $H\sim h^{-1}$ and is well modelled by 1st order cnoidal shoaling, see Figure 1.5.5. As a simple practical approach, one may add the empirical finite wave height correction

$\left(1+\frac{3}{8}(\frac{H_o}{L_o})^{1.5}(k_oh)^{-3}\right)$ to the small amplitude result (1.7.6), i e:

$$\frac{H}{H_o} = \frac{1}{\sqrt[4]{4k_oh}}\left(1+\frac{1}{4}k_oh+\frac{13}{228}(k_oh)^2\right)\left(1+\frac{3}{8}(\frac{H_o}{L_o})^{1.5}(k_oh)^{-3}\right) \qquad (1.7.7)$$

the accuracy of which is illustrated by the comparison with experiments in Figure 1.7.4

1.7.3   Refraction

When waves encounter a depth change the speed of propagation will change and the direction of propagation will change as indicated by the examples in Figure 1.7.5.

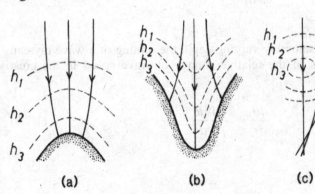

(a)   (b)   (c)

Figure 1.7.5: Refraction patterns caused by the change in wave speed with changing depth. After Svendsen and Jonsson (1976).

(a) Concentration of the wave energy around a headland.
(b) Dilution of the wave energy in a bay.
(c) Focusing of waves by a shoal.

    Refraction can also occur when waves change their absolute speed because they are travelling on a non-uniform current. This process occurs near river entrances and where ocean currents squeeze around a corner of a continent. It is discussed extensively by Jonsson (1990).
    The refraction process can be understood in terms of the turning of the wave in Figure 1.7.6 as it experiences a change of speed in passing over a straight step into a shallower depth.

Figure 1.7.6: Due to the change of celerity, the angle α between wave crest and the line of depth change will change like the direction of light entering a prism.

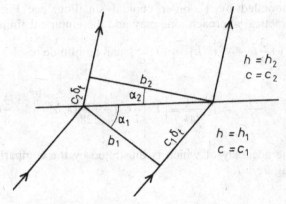

From the two right angled triangles in Figure 1.7.6 it is seen that

$$\frac{\sin \alpha_2}{\sin \alpha_1} = \frac{c_2}{c_1} \qquad (1.7.8)$$

which is called *Snell's law*.
    For arbitrary, smoothly varying depth the turning of a wave ray can be expressed in terms of the relative change of wave speed in the crest direction

$$\frac{d\theta}{ds} = -\frac{1}{c}\frac{dc}{dn} \qquad (1.7.9)$$

see Figure 1.7.7.

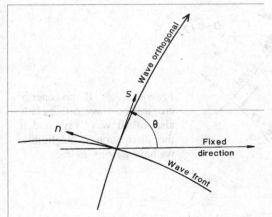

Figure 1.7.7: If c increases in the n-direction the angle θ will decrease. The effect is similar to the turning of a car that pulls more on the left wheel than on the right.

In order to propagate wave rays over topography it is thus necessary to compute c and its derivative in each step. This is made easier by using the explicit approximation to the sine wave expression for c:

$$c = c_o \tanh kh \approx c_o \sqrt{k_o h}(1 - \frac{1}{6}k_o h) \qquad (1.4.24)$$

which is within 1% for $k_o h < 1.62$. By differentiating and inserting into (1.7.8) this leads to the explicit "ray turning equation"

$$\frac{d\theta}{ds} = -\frac{\frac{1}{2} - \frac{3}{12}k_o h}{k_o h - \frac{1}{6}(k_o h)^2} k_o \frac{dh}{dn} \qquad (1.7.10)$$

1.7.4 Combined refraction and shoaling

The combined effect of refraction and shoaling is a change of wave height, which can be analysed fairly simply in accordance with Figure 1.7.8 under the assumptions that no energy crosses the wave orthogonals and that energy dissipation is negligible.

Conservation of energy flux between the orthogonals yields

$$E_{f1}b_1 = E_{f2}b_2 \qquad (1.7.11)$$

and since $E_f \propto H^2 c_g$ the relation between the wave heights is

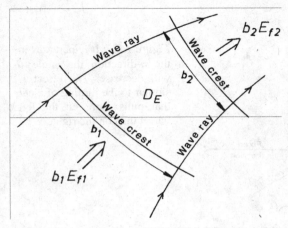

Figure 1.7.8: If no energy crosses the wave orthogonals also called wave rays and if the dissipation is negligible, we have $E_{f1}b_1 = E_{f2}b_2$.

$$\frac{H_2}{H_1} = \sqrt{\frac{c_{g1}}{c_{g2}}\frac{b_1}{b_2}} \qquad (1.7.12)$$

When some of the computed wave rays intersect or get very close, see Figure 1.7.5c or Figure 1.7.10, this relationship becomes invalid. In such cases the wave height gradient in the direction of the wave crest will be very large and the wave energy will actually spread away from the points of very high energy density by the process of diffraction, which will be described in Section 1.8.

1.7.5 Sine waves on straight and parallel bottom contours

For the special case of straight and parallel depth contours the lengths b_1 and b_2 can be expressed as a function of the depths h_1 and h_2 only, simply by using Snell's law. From Figure 1.7.6 it follows that

$$\frac{b_2}{b_1} = \frac{\cos\alpha_2}{\cos\alpha_1} \qquad (1.7.13)$$

and since $\cos^2 x \equiv 1-\sin^2 x$ and from Snell's law $\sin\alpha_2 = \frac{c_2}{c_1}\sin\alpha_1$ we have

$$\frac{b_2}{b_1} = \frac{\sqrt{1-\frac{c_2^2}{c_1^2}\sin^2\alpha_1}}{\cos\alpha_1} \qquad (1.7.14)$$

78

For comparison with deep water we use the notation $H(h) = K_s K_r H_o$ in analogy with pure shoaling. K_r is called the refraction coefficient. We find

$$H = K_s K_r H_o = K_s \sqrt{\frac{b_o}{b}} H_o = K_s \sqrt{\frac{\cos \alpha_o}{\sqrt{1 - (\frac{c}{c_o})^2 \sin^2 \alpha_o}}} H_o \quad (1.7.15)$$

which assumes straight and parallel bottom contours.

Again, for calculations based on sine wave theory it is reasonable to use explicit approximations. For shallow water, i e, most surf zone problems the simple expression

$$K_r = \sqrt{\cos \alpha_o} \quad (1.7.16)$$

may be used. For greater depths more terms are needed as indicated by Figure 1.7.9 which shows the 1% error limits for the three formulae (1.7.16), (1.7.17) and (1.7.18) in the $(k_o h, \alpha_o)$-plane.

$$K_r = \sqrt{\cos \alpha_o} (1 + \frac{1}{4} k_o h \sin^2 \alpha_o) \quad (1.7.17)$$

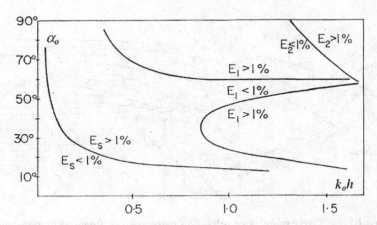

Figure 1.7.9: 1% error limits for the explicit approximations to the sine wave refraction coefficient (1.7.15): $E_s \sim (1.7.16)$, $E_1 \sim (1.7.17)$ and $E_2 \sim (1.7.18)$. Valid only for straight parallel bottom contours.

$$K_r = \frac{\sqrt{\cos \alpha_o}}{\sqrt[4]{1 - k_o h (1 - \frac{1}{6} k_o h)^2 \sin^2 \alpha_o}} \qquad (1.7.18)$$

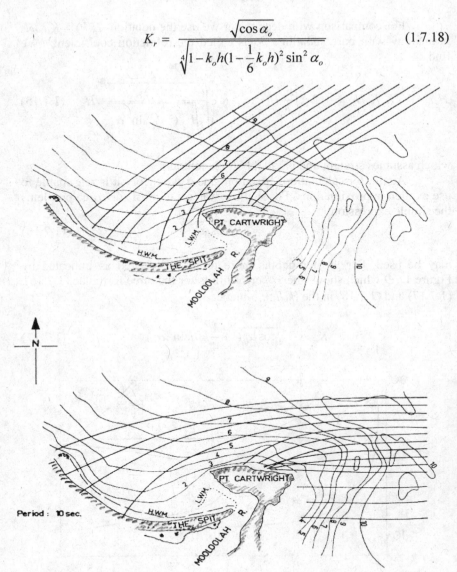

Figure 1.7.10: Two examples of refraction patterns over a natural topography. We see that the refraction pattern depends strongly on the offshore wave direction. It will also in general depend on the wave period. As mentioned above, the crossing orthogonals are an artifact of the simplified analysis which neglects diffraction. Refraction analysis by Dr Michael Gourlay.

1.8   DIFFRACTION OF WATER WAVES

Diffraction is the process by which waves swing around the ends of breakwaters or small islands, as in Figure 1.7.2, and spread out behind gaps that are only a few wavelengths wide. It happens because energy will travel along (as well as perpendicular to) the wave crest if the wave height varies along the crest.

Diffraction can be understood in terms of *Huygens' principle*:

Every point of a wave front is the source of circular wavelets with the same period and speed as the original wave.

An example of the corresponding construction of wave fronts near an obstacle is shown in Figure 1.8.1.

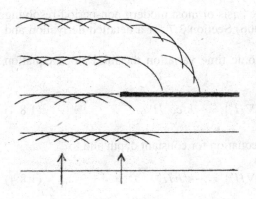

Figure 1.8.1: Construction using Huygens' principle of the wave fronts near the end of a breakwater for constant depth.

Huygens' principle does however not give the wave height distribution along the wave crests. At constant depth, it is governed by the Helmholtz equation for the complex wave height H'

$$\frac{\partial^2 H'}{\partial x^2} + \frac{\partial^2 H'}{\partial y^2} = -k^2 H' \qquad (1.8.1)$$

under the assumptions of constant wave length and no energy dissipation. The complex wave height H' is the usual wave height times a phase function of (x, y) in general. For a simple wave propagating in the x-direction only:

$$\eta(x,t) = \frac{H}{2}e^{i(\omega t - kx)} \tag{1.8.2}$$

it is

$$H' = H'(x) = He^{-ikx}. \tag{1.8.3}$$

The solution to (1.8.1) depends heavily on the amount of reflection returned from the breakwater and other boundaries.

The Shore Protection Manual and the Coastal Engineering Manual contain diagrams for estimation of wave heights due to diffraction around various breakwater geometries under the assumption of constant water depth.

For slowly varying depth, diffraction as well as refraction of sine waves can be accounted for through the so-called Mild Slope Equation, which was derived by Berkhof (1972).

In various forms it is the basis of most modern non-period-resolving wave models, see Svendsen (2006) Section 3.7 for a detailed derivation and discussion.

Assuming simple harmonic time variation the mild slope equation for H' reads

$$\nabla\left(cc_g\nabla H'\right) = -k^2 cc_g H' \tag{1.8.4}$$

which reduces to the Helmholz equation for constant depth and to

$$\nabla\left(h\nabla H'\right) = -k^2 h H' \tag{1.8.4}$$

for long waves. The ∇-operator is defined by $\nabla H' = \dfrac{\partial H'}{\partial x} + \dfrac{\partial H'}{\partial y}$.

1.9 REFLECTION AND SEICHES

Beaches, breakwaters and sea walls reflect the incoming waves to varying extents depending on the wave length, the slope steepness and the porosity and surface roughness of the slope. In the case of submerged and/or very porous structures, some of the wave energy may be transmitted to water on the other side. The wave energy which is neither reflected nor transmitted is dissipated into heat.

The ratios of the heights of the reflected wave (H_r) and the transmitted wave (H_t) to that of the incoming wave H_i are the reflection coefficient $K_r = H_r/H_i$ and the transmission coefficient $K_t = H_t/H_i$.

In enclosed or partly enclosed basins with reflecting sides pulsating forcing with a suitable frequency can set up standing wave oscillations with wave lengths of the order of the basin dimensions. These oscillations are called seiches.

1.9.1 Superposition of reflected and incoming waves

In front of a reflecting structure or beach, the reflected waves will be superimposed on the incoming waves and the amplitude of the total motion will vary with distance from the reflecting boundary because the two waves will enhance respectively cancel each other depending on the local phase difference. At the boundary, the two are in phase and the resulting surface elevation range is $H_i + H_r = H_i(1 + K_r)$. At a distance of $L/4$ from the wall the reflected wave will have traveled $L/2$ further than the incoming wave and hence will be $180°$ out of phase, see Figure 1.9.1.

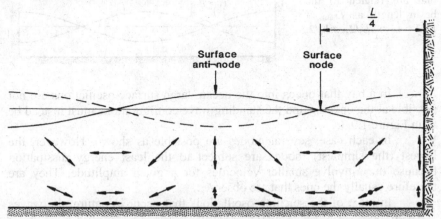

Figure 1.9.1: Surface elevations and velocity field due to total reflection of sine waves from a wall that is parallel to the crests of the incoming waves.

Here the surface motion of the reflected wave therefore partly or completely cancels that of the incoming wave. The surface elevation range is $H_i-H_r = H_i(1-K_r)$.

The velocity field under the "standing waves" in Figure 1.9.1 is very different from that under progressive waves. While u and w are $90°$ out of phase in progressive waves they are everywhere in phase in the standing waves. At the wall, the horizontal velocities vanish while the vertical velocities are maximum. Under the surface nodes, the vertical velocities vanish while the horizontal velocities are maximum.

For a smooth, impermeable, vertical wall one can obtain virtually perfect reflection. If the slope is rough some energy will be lost to surface friction and if the slope is permeable further energy will be lost to friction by the flow inside the breakwater.

1.9.2   Seiches

In an enclosed basin with reflecting end walls standing wave oscillations can be generated by pulsating forcing with a suitable frequency, see Figure 1.9.2.

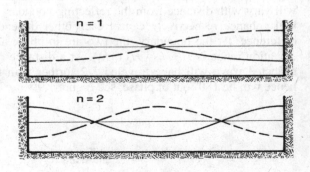

Figure 1.9.2: Seiches in an enclosed basin correspond to standing waves with wave lengths that are related to the basin length as $L_{basin} = L_1/2 = L_2 = 3L_3/2...nL_n/2$.

In a bay that opens into the ocean, large surface oscillations are not possible at the opening and the standing wave configurations must instead be as in Figure 1.9.3.

In each case, several modes are possible as shown. However, the lowest (the simplest) modes are subject to the least energy dissipation because they involve smaller velocities for a given amplitude. They are therefore usually the ones that are observed.

In most of the seiching oscillations that occur in nature, the waves are shallow water waves so that $L_n = T_n\sqrt{gh}$.

84

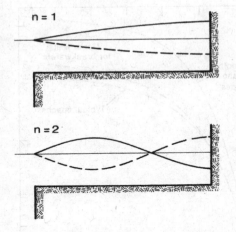

Figure 1.9.3: Seiches in a bay are standing waves with possible wave lengths L_n that are related to the bay length $L_{bay} = L_1/4 = 3L_2/4 = 5L_3/4 \ldots (2n+1)L_n/4$.

On length scales of the order one kilometre or less the forcing is usually wind gusts or surf beat with periods up to a few minutes. On a much larger scale, the forcing agent can be the variation in atmospheric pressure or general wind direction due to moving weather systems. Such oscillations in the Adriatic Sea are the most frequent causes for flooding in the city of Venice.

In a square basin, resonant oscillations can be set up in both directions of the sides and in the diagonal directions. The two types of oscillations will have different periods. Similarly, harbour basins with more or less complicated plan shape can suffer from seiches with many different periods. The most common forcing for such harbour oscillations are long waves generated by the wind-wave groups having periods in the range 50s to 300s.

In order to avoid unwanted ship movements due to these seiches, as many as possible of the basin walls should be designed to absorb rather than reflect the wave energy. A comprehensive treatment of seiches and harbour resonance can be found in Raichlen (1966).

1.9.3 Reflection from a slope

If the slope of the wall is decreased from vertical the reflection coefficient will initially remain close to unity. Eventually however, the waves will start dissipate energy due to breaking. Consequently, the reflection coefficient will decrease as a function of the *surf similarity parameter* $\xi = \tan \beta \sqrt{L_o / H_i}$:

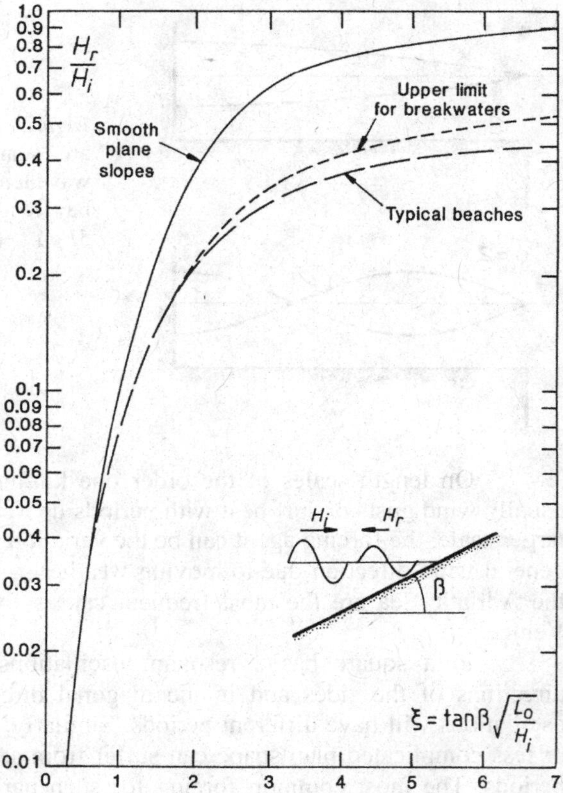

Figure 1.9.4:
Reflection coefficients as function of the surf similarity parameter. After the Shore Protection Manual.

$$K_r = K_r(\tan \beta \sqrt{L_o / H_i}) \qquad (1.9.1)$$

see Figure 1.9.4.

1.9.4 Transmission and reflection at a depth change

Similarly to light being partially reflected from a glass surface where it is forced to change its speed of propagation, water waves will be partially reflected at a depth change which occurs over a short distance compared with the wave length, see Figure 1.9.5.

In general the energy fluxes due to the reflected and the transmitted waves must add up to that of the incoming wave minus any energy

86

dissipation that may occur at the "step". Hence, with negligible dissipation we have

$$E_{fr} + E_{ft} = E_{fi} \qquad (1.9.2)$$

or cancelling the factor $\rho g H_i^2/8$ in the expression (1.4.38) for energy flux in sine waves

$$K_r^2 c_{g1} + K_t^2 c_{g2} = c_{g1} \qquad (1.9.3)$$

Another matching condition can be obtained by stipulating that the discharges Q_i, Q_r and Q_t corresponding to the three waves satisfy

$$Q_i(x_{\text{step}}, t) - Q_r(x_{\text{step}}, t) \equiv Q_t(x_{\text{step}}, t) \qquad (1.9.4)$$

This can only be satisfied if the discharge amplitudes satisfy the corresponding

$$|Q_i| - |Q_r| \equiv |Q_t| \qquad (1.9.5)$$

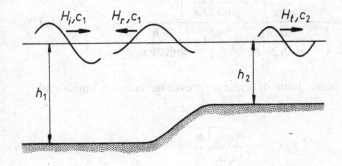

Figure 1.9.5: An abrupt depth change will cause reflection of part of the wave energy.

To implement this flux condition for sine waves let us first calculate the discharge amplitude for a sine wave

$$|Q| \approx \int_{-h}^{o} u_{\max} dz = \omega \frac{H}{2} \frac{1}{\sinh kh} \int_{-h}^{o} \cosh k(z+h) \, dz = \frac{\omega}{k} \frac{H}{2} = c \frac{H}{2} \qquad (1.9.6)$$

The initial approximation is to integrate only to $z=0$ instead of $z=\eta$ and we have used $\int \cosh k(z+h)\,dz = \dfrac{1}{k}\sinh k(z+h)$ and the identity $\dfrac{\omega}{k} = c$.

Inserting this expression into (1.9.5) leads to

$$(1-K_r)c_1 = K_t c_2 \tag{1.9.7}$$

and solving together with (1.9.3) for the reflection coefficient and the transmission coefficient then leads to

$$K_r = \frac{\left(1+\dfrac{2k_2h_2}{\sinh 2k_2h_2}\right)c_1 - \left(1+\dfrac{2k_1h_1}{\sinh 2k_1h_1}\right)c_2}{\left(1+\dfrac{2k_2h_2}{\sinh 2k_2h_2}\right)c_1 + \left(1+\dfrac{2k_1h_1}{\sinh 2k_1h_1}\right)c_2} \tag{1.9.8}$$

and

$$K_t = \frac{2\left(1+\dfrac{2k_1h_1}{\sinh 2k_1h_1}\right)c_1}{\left(1+\dfrac{2k_2h_2}{\sinh 2k_2h_2}\right)c_1 + \left(1+\dfrac{2k_1h_1}{\sinh 2k_1h_1}\right)c_2} \tag{1.9.9}$$

In shallow water both of these expressions become considerably simpler:

$$K_r = \frac{1-\sqrt{h_2/h_1}}{1+\sqrt{h_2/h_1}} \tag{1.9.10}$$

and

$$K_t = \frac{2}{1+\sqrt{h_2/h_1}} \tag{1.9.11}$$

If the waves encounter a step at an angle as in Figure 1.9.6 the energy flux condition, considering energy flux across the step becomes

$$E_{fi}\cos\alpha_1 - E_{fr}\cos\alpha_1 = E_{ft}\cos\alpha_2 \qquad (1.9.12)$$

For the case of linear shallow water waves it turns out that the three conservation laws for respectively shore-normal mass flux, longshore momentum flux and shorenormal energy flux all boil down to

$$H_i^2\cos\alpha_1 = H_t^2 \sqrt{\frac{h_2}{h_1}}\sqrt{1 - \frac{h_2}{h_1}\sin^2\alpha_1} + H_r^2\cos\alpha_1 \qquad (1.9.13)$$

when Snell's law delivers $\dfrac{\sin\alpha_2}{\sin\alpha_1} = \sqrt{\dfrac{h_2}{h_1}}$ and $\cos\alpha_2 = \sqrt{1 - \dfrac{h_2}{h_1}\sin^2\alpha_1}$.

We note that for big depth increases, corresponding to $\dfrac{h_2}{h_1}\sin^2\alpha_1 > 1$ there are no real-valued solutions for $\cos\alpha_2$. In that case no waves propagate across the step.

A second independent equation is obtained if surface continuity is required:

$$H_t = H_i \pm H_r \qquad (1.9.14)$$

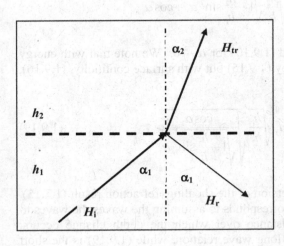

Figure 1.9.6: Refraction and reflection of waves which encounter a depth change at an angle.

In terms of the transmission coefficient $K_t = H_t/H_i$ and the reflection coefficient $K_r = H_r/H_i$ these two equations become

$$\cos\alpha_1 = K_t^2 \sqrt{\frac{h_2}{h_1}} \sqrt{1 - \frac{h_2}{h_1}\sin^2\alpha_1} + K_r^2 \cos\alpha_1 \qquad (1.9.15)$$

and

$$K_t = 1 + K_r \qquad (1.9.16)$$

which have the solution

$$K_t = \frac{2\cos\alpha_1}{\sqrt{\frac{h_2}{h_1}}\sqrt{1 - \frac{h_2}{h_1}\sin^2\alpha_1} + \cos\alpha_1} \qquad (1.9.17)$$

and

$$K_r = \frac{\cos\alpha_1 - \sqrt{\frac{h_2}{h_1}}\sqrt{1 - \frac{h_2}{h_1}\sin^2\alpha_1}}{\sqrt{\frac{h_2}{h_1}}\sqrt{1 - \frac{h_2}{h_1}\sin^2\alpha_1} + \cos\alpha_1} \qquad (1.9.18)$$

which agree with (1.9.11) and (1.9.10) for $\alpha_1 \to 0$. We note that with energy flux conservation expressed by (1.9.15) but with surface continuity, (1.9.16), relaxed we find

$$H_t = H_i \sqrt[4]{\frac{h_1}{h_2}} \sqrt{\frac{\cos\alpha_1}{\sqrt{1 - \frac{h_2}{h_1}\sin^2\alpha_1}}} \qquad (1.9.19)$$

which is the shallow water version of the shoaling+refraction result (1.7.15). Invoking surface continuity corresponds to assuming the waves the waves to be long compared to the distance over which the depth change occurs. Equation (1.9.17) is thus the long wave relation, while (1.9.19) is the short wave relation. The two are compared in Figure 1.9.7.

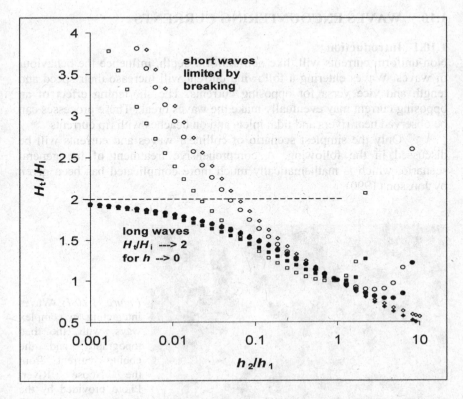

Figure 1.9.7: Transmission coefficients K_t, ♦ ● ■, for long waves and $\alpha_1 = 0°$, $20°$ and $40°$ respectively. Compared with the short wave result for the same incidence angles, ◊ ○ □. The long waves end up smaller because they are partially reflected as expressed by (1.9.18). The heights of the transmitted short waves will eventually be limited by breaking.

1.10 WAVES ENCOUNTERING CURRENTS

1.10.1 Introduction

Non-uniform currents will, like non-uniform depth, influence the behaviour of waves. Waves entering a following current will increase their speed and length and vice versa for opposing currents. The shortening effect of an opposing current may eventually make the waves break. These processes can be observed near rivers and tidal inlets and on beaches with rip currents.

Only the simplest scenario of collinear waves and currents will be discussed in the following. A comprehensive treatment of the general scenario, which is mathematically much more complicated has been given by Jonsson (1990).

Figure 1.10.1: Waves interacting in complex ways with the bed topography and the ebbing current from the Noosa River. Photo provided by the State of Queensland, Department of Natural Resources and Water [1972].

1.10.2 Dispersion relation for waves on a current

A steady and uniform current does not influence the behaviour of waves in the sense that, in a frame of reference, which follows that current (the boatman in Figure 1.10.2), the behaviour of the waves is exactly as in the no-current-situation. Hence, for a given sine wave with period T_r in the frame of reference following the current, the wave number is determined by the usual dispersion relation (1.4.13):

$$kh \tanh kh = \frac{4\pi^2}{gT_r^2} h = \frac{\omega_r^2}{g} h \qquad (1.10.1)$$

Conversely, the relative period for a sine wave with length L is, as usual

$$T_r = \sqrt{\frac{2\pi L}{g \tanh(\frac{2\pi}{L} h)}} \qquad (1.10.2)$$

and the corresponding celerity $c_r = L/T_r$. We use subscripts r (~ relative), to refer to such a frame of reference.

The trouble is that an observer who does not follow the current sees a different wave period. This is easiest explored in terms of the wave

Figure 1.10.2: Relative and absolute wave properties. Artwork by Dr Kevin Bodge.

celerities seen by the different observers. To an observer who is at rest relative to the ground (the observer in the tree in Figure 1.10.2), the wave celerity is

$$c_a = c_r + U_a \qquad (1.10.3)$$

where U_a is the current speed and subscript a refers to the absolute frame of reference. Since $c = \omega/k$ in any frame of reference, Equation (1.10.3) gives the relation between the angular frequencies

$$\frac{\omega_a}{k} = \frac{\omega_r}{k} + U_a \qquad (1.10.4)$$

and hence

$$\omega_r = \omega_a - kU_a \qquad (1.10.5)$$

With this, the dispersion relation (1.10.1) can be rewritten in terms of the absolute angular frequency. We find

$$gk \tanh kh = (\omega_a - kU_a)^2 \qquad (1.10.6)$$

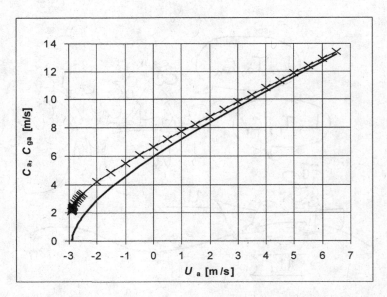

Figure 1.10.3: Absolute celerity c_a -×-×-×- and group velocity c_{ga} ——— as function of current speed U_a for $T_a = 8\text{s}$, $h = 5\text{m}$.

94

For sine waves propagating in constant depth from an area without current into collinear currents of variable speed U_a (positive in the direction of wave propagation) the variation of the absolute celerity and group velocity are shown in Figure 1.10.3.

We see that on very strong following currents, the waves become very long and we get the shallow water relation: $c_{ga} \approx c_a$.

A strong enough opposing current will prevent the waves from propagating upstream: $c_{ga} \rightarrow 0$. This phenomenon is called *wave blocking*.

Example 1.10.1:
Find the celerity and group velocity for sine waves with period $T_a = 8s$, at $h = 5m$ propagating against a current, $U_a = -2m/s$.

Solving the dispersion relation (1.10.6) gives $k = 0.1872m^{-1}$, corresponding to $L = 33.6m$ and, through (1.10.2), this gives $T_r = 5.42s$.

Using normal sine wave procedures based on T_r then gives $c_r = 6.20m/s$ and $c_{gr} = 4.93m/s$. Finally the absolute speeds are obtained by adding U_a: $c_a = 4.20m/s$ and $c_{ga} = 2.93m/s$.

☺

1.10.3 Wave height variation due to varying current speed
Solving the dispersion relation (1.10.6) and applying the relevant sine wave formulae enables the calculation of sine wave quantities like k, L, c_r, c_a c_{gr}, and c_{ga}, but not immediately the wave height. The wave height will however, in analogy with shoaling, change on a varying current. For the simple case of collinear currents and no dissipation one gets

$$\frac{d}{dx}\left([1+\frac{U_a}{c_r}][U_a + c_{gr}]E \right) = 0 \qquad (1.10.7)$$

Equation (1.10.7) is a special case of the Wave Action Conservation Equation, cf Jonsson (1990), which handles 2D-horizontal wave height variation on variable depth and currents, with dissipation, as long as diffraction is insignificant. *Action* is, in relation to waves on a current, defined as

$$A = \frac{E}{\omega_r} = \frac{\frac{1}{8}\rho g H^2}{\omega_r} \qquad (1.10.8)$$

The fact that a simpler conservation law can be formulated for action than for energy has analogies in other fields of physics. For example, a photon, which is reflected from a moving mirror, will not conserve its energy (= frequency × Planck's Constant) because its frequency changes. Its action (= Planck's Constant) is however conserved.

Example 1.10.2:

Find the wave height variation $H(x)$ for sine waves propagating towards a river entrance at $x=0$, seaward of which the depth varies as $h_E - S\,x$ (x positive in the onshore direction), where the entrance depth is h_E and S is the bed slope. The current speed is $U_a(x) = \pm U_E / [(1 - x/2B)\,h/h_E]$ where B is the width of the river mouth. The signs + and − apply for flooding and ebbing currents respectively.

We apply Equation (1.10.7) in the form

$$\left([1 + \frac{U_a(x)}{c_r(x)}][U_a(x) + c_{gr}(x)]\frac{1}{8}\rho g H^2(x) \right) = \left([1 + \frac{U_{a,o}}{c_{ro}}][U_{a,o} + c_{gro}]\frac{1}{8}\rho g H_o^2 \right) \quad (1.10.9)$$

where, as usual, subscript "o" denotes deep-water properties.

Since the current vanishes in deep water the contents of the right hand side becomes simply $\frac{1}{2}c_o\,\frac{1}{8}\rho g H_o^2$ so, the wave height is determined from

$$H(x) = H_o \sqrt{\frac{\frac{1}{2}c_o}{[1 + \frac{U_a(x)}{c_r(x)}][U_a(x) + c_{gr}(x)]}} \quad (1.10.10)$$

which becomes the pure-shoaling-formula (1.7.5) for zero current.

For the special case of $(T_a, H_o, B, h_E, U_E) = (10\text{s}, 2\text{m}, 80\text{m}, 3\text{m}, 2\text{m/s})$, the results of solving the dispersion relation (1.10.6) to find local values of k, c_r, c_{gr} and then applying (1.10.10) to find the wave heights are shown in Figure 1.10.4.

☺

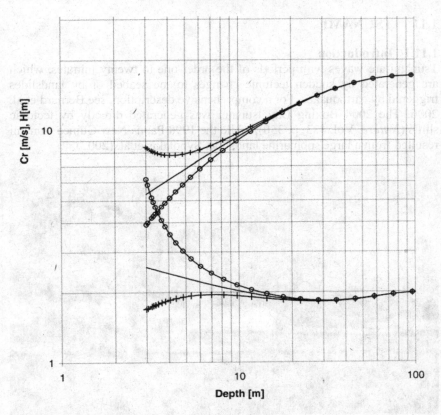

Figure 1.10.4: Variation of c_r upper curves and H, lower curves for waves propagating towards a river entrance at depth 3m. Plain lines correspond to no current, + to a flooding tide and ○ to ebbing tide.

Figure 1.10.4 shows how the current affects the relative wave speed as well as the absolute wave speed. This is because the waves get stretched, respectively shortened by the currents.

The very steep wave height increase on the ebbing current will of course be truncated by wave breaking before H becomes equal to h.

1.11 TSUNAMI

1.11.1 Introduction

Tsunami are waves with periods of the order one to twenty minutes, which are generated by sudden tectonic changes to the seabed or by landslides triggered by earthquakes. For a comprehensive description, see Bernard et al. 2006. The 2004 Boxing Day tsunami was generated directly by tectonic shifts (*Nature*, Vol. 433, p. 350), while the 1998 Papua New Guinea tsunami resulted from a large submarine landslide, Synolakis et al. (2002).

Figure 1.11.1: Sign on the Fuji Coast in Japan warning people about the risk of tsunami inundation in case of seismic activity. Catfish have traditionally been used for earthquake warnings since they change their behaviour when they sense earth tremors.

The word tsunami (same in plural and singular) is Japanese and indeed, a lot of the groundbreaking work on tsunami has been done in Japan. Tsunami rank highly among natural disasters with respect to deadly destructive power, e g, the tsunami resulting from the eruption of Santorini around 1470BC has been held responsible for the disappearance of the Minoan civilisation and the 2004 Boxing Day tsunami which hit the shores of the NE Indian Ocean after an earthquake off the coast of Sumatra, killed more than 200,000 people. This enormous destructive power is perhaps surprising since tsunami wave heights rarely exceed 1m in deep water. However, they shoal to many times the deepwater height as they move into shallow water. They may or may not break and form bores as they reach the shore. In either case they may run up to heights exceeding 50m in extreme cases. Bryant (2008) gives a comprehensive record of historical events. A couple of curious features of tsunamis are: 1st, often the first sign of an impending tsunami is a drop in water level, i e, many tsunamis spread out with a leading trough rather than a leading crest. 2nd, when a tsunami approaches a small island, it may refract and diffract around in such a way that the two waves combining at the back may yield a combined height which is greater than that experienced on the side facing the original tsunami.

1.11.2 Generation of tsunami by tectonic uplift

Because the typical oscillation frequencies of earthquakes are of the order 1Hz, i e, hundreds of times faster than tsunami frequencies, earthquakes do not generate tsunamis in the way that a wave paddle generates same-frequency waves in a laboratory wave flume. However, if the earthquake generates lasting seabed dislocations, these will be translated into a water surface deformation, which will subsequently travel in all directions more or less in accordance with shallow water wave theory. The tsunami period is then derived from the horizontal scale of the surface/seabed deformation through the dispersion relation (1.4.11).

The translation of the sea bed deformations into initial sea surface deformations is analogous to the way in which water surface elevations are translated into seabed pressures as expressed by Equation (1.4.33). That is, if the seabed is instantaneously deformed into a sine wave with length L and height H_{bed}, the surface at depth h above is transformed into a sine wave with the same length and with height

$$H_{surface} = \frac{H_{bed}}{\cosh k_{bed} h} \tag{1.11.1}$$

where $k_{bed} = 2\pi/L$.

If the bed deformation $\zeta(x)$ is not a sine curve, but is assigned local wave numbers by the rule $k^2 = -\zeta_{xx}/\zeta$, the smoothing of the shape by the water column can be estimated by

$$\eta = \frac{\varsigma}{\cosh\left(\sqrt{-\varsigma_{xx}/\varsigma}\, h\right)} = \varsigma + \frac{h^2}{2}\varsigma_{xx} + \dots \tag{1.11.2}$$

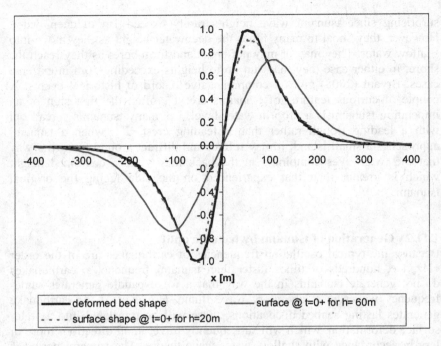

Figure 1.11.2: Seabed deformation and resulting, initial sea surface shapes for h=20m and h=60m. As indicated by (1.11.1) the sea surface amplitude is reduced with greater depth. In addition, for a non-sine-wave shape, greater depth generally leads to a longer wavelength, as higher wave numbers are dampened more.

An example is shown in Figure 1.11.2.

The way in which the sea surface subsequently develops is evaluated by propagating the Fourier components of the initial shape as free waves in either direction. Conservation of momentum applies after the tectonic forcing has stopped. If it did not impart horizontal momentum to the water, the pattern of spreading free waves should be symmetric.

1.11.3 Tsunami generation by gradual sea bed deformations

The speed with which the bed deformation happens or the speed of the landslide influences the size and shape of the tsunami. If a landslide moves with speeds much less than $\sqrt{gh}$, the water surface adjusts through states of quasi equilibrium and little wave motion results. If, on the other hand, $V > \sqrt{gh}$, steep shock waves (bores) may be generated which dissipate energy by breaking. At intermediate values of $V/\sqrt{gh}$ wave trains of different shapes are generated, which may travel without breaking but constantly changing form because they generally consist of a mixture of forced and free waves with different speeds. Instructive examples can be found in the laboratory study of Fritz et al. (2004). They considered scenarios like Figure 1.11.3.

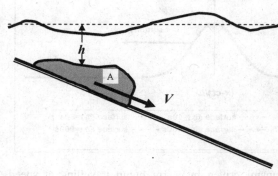

Figure *1.11.3*: Landslide of cross sectional area A moving down slope at speed V under water at depth h.

1.11.4 Tsunami generated by a hump sliding on a horizontal sea floor

The simple case of tsunami generation by hump $z_b(x,t) = Z_o f(x-ct)$ sliding along a horizontal sea floor can be treated in analogy with other forced long waves as shown in Section 1.3.3, where we found the steady forced wave solution $A_{forced}f(x-ct)$ with A_{forced} given by

$$A_{forced} = \frac{c^2 Z_o}{c^2 - gh} \qquad (1.3.21)$$

This is all that is seen over a hump, which has been moving with constant speed for so long that all transient or free waves have moved away.

The transient or initial growth stages are, as detailed in Section 1.3.3, modelled by seeing the initial flat sea surface $\eta(x,0)$ as the superposition of the steady forced wave $\eta_{forced}(x,0) = A_{forced}f(x-ct)$ and two free waves: The

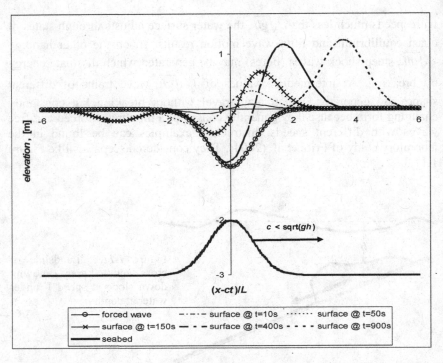

Figure 1.11.4: Developing tsunami driven by a 1m hump travelling at speed c=10m/s while the free-wave speed is $\sqrt{gh}$ =14m/s. A_{forced}=−1.04m, $(A_{\text{free}}^{+}, A_{\text{free}}^{-})$ = (1.25m, −0.21m). The coordinate system follows the forced wave and hence the hump which is really at depth 20m, but drawn at −3m for illustration purposes. At t=150s $\eta^{-}(x,t)$ is still visible on the left. At t=900s $\eta^{-}(x,t)$ has disappeared while $\eta^{+}(x,t)$ is about to become separated from the forced wave.

forward moving $\eta^{+}(x,t)=A_{\text{free}}^{+}f(x-\sqrt{gh}t)$ and the backward moving $\eta^{-}(x,t) = A_{\text{free}}^{-}f(x+\sqrt{gh}t)$. An example is shown in Figure 1.11.4.

For the scenario in Figure 1.11.4, the free wave amplitudes are determined by $A_{\text{free}}^{+} + A_{\text{free}}^{-} + A_{\text{forced}} = 0$, which corresponds to the initially flat ocean surface: $\eta_{\text{forced}}(x,0)+\eta^{+}(x,0)+\eta^{-}(x,0)$ =0 and the zero flow condition

$$-H_o c + A_{\text{free}}^{+}\sqrt{gh} - A_{\text{free}}^{-}\sqrt{gh} + A_{\text{forced}}c = 0 \qquad (1.11.3)$$

where the first term corresponds to the return flow generated by the hump sliding under a flat surface. With A_{forced} given by (1.3.21), these lead to

$$\left(\frac{A_{\text{free}}^{+}}{Z_o}, \frac{A_{\text{free}}^{-}}{Z_o} \right) = \left(\frac{1}{2} \frac{\dfrac{c}{\sqrt{gh}}}{1 - \dfrac{c}{\sqrt{gh}}}, \; -\frac{1}{2} \frac{\dfrac{c}{\sqrt{gh}}}{1 + \dfrac{c}{\sqrt{gh}}} \right) \qquad (1.11.4)$$

These show that the backward moving free wave is always a negative surge while the sign of the forward moving free wave changes at "resonance", $c = \sqrt{gh}$. Extensions of this simple long-wave solution has been given by Tinti et al. (several papers).

At resonance, the linear 1D long-wave solution is a linearly ($\sim t$) growing wave in the shape of $\dfrac{\partial f}{\partial x}$ where $f(x - \sqrt{ght})$ is the shape of the forcing, i e, the Gaussian hump in Figure 1.11.4 generates an N-wave if it travels at $c = \sqrt{gh}$, see Section 1.3.3.5. The character of the resonant solution is completely changed by the combination of non-linear and non-shallow-water effects as described by Grimshaw et al. (2007).

1.11.5 Shoaling, breaking and runup

Details of the far field behaviour and shoaling and runup can be found in Sato (1996) and in Lynett and Liu (2005). Some examples are shown in Figure 1.11.5.

We note that, for all but the smallest amplitude, the wave front breaks up into shorter waves. This behaviour cannot be modelled by the non-dispersive nonlinear shallow water equations (Section 1.5.2), but Sato (1996) managed to model it quite well with a Boussinesq wave model. Field observations of such short waves with periods around 10s was reported by Shuto (1985) and were prominent in the television footage of the Boxing Day 2004 tsunami.

Secondly, we note that the surge level continues to rise after the passage of the front. A feature, which seems to be common for tsunami but, which is rarely observed with bores generated by wind waves breaking on beaches, see Figure 2.4.1. The sustained rise after the arrival of the front is even more pronounced at the still water shoreline as shown in Figure 1.11.6.

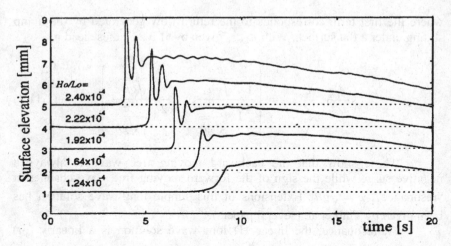

Figure 1.11.5: Laboratory measurements of "tsunami" surface shapes close to the shoreline (at still water depth 5.63cm). Sine waves with T=40s (L_o=2500m) were generated in the deep (h=1.0m) section of a 163m long flume and propagated along a long 1/200 slope after a short 1/10 rise to h=0.5m. Data from Tsuruya et al. (1984) reported by Sato (1996).

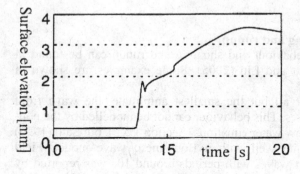

Figure 1.11.6: Time series of the water level at the still water shoreline. H_o/L_o=1.92×10^{-4}. Data from Tsuruya et al. (1984) reported by Sato (1996).

By comparison with Figure 1.11.5 it can be seen that the short waves at the front have decayed by breaking while the main wave has continued to grow by shoaling between still water depth 5.63cm and the still water shoreline.

1.11.6 Inundation from tsunami

Inundation from tsunami is clearly one of the most severe types of natural disaster. More than 200,000 people were killed by the 2004 Boxing Day tsunami. Long term damage resulted from sea water entering drinking water wells and agricultural land, contaminating both with salt which may take decades to get flushed out.

Accurate field measurements of local water depth and/or fluid velocities are rare, and the difference in scale and the presence of roughness (houses and trees) in the real scenarios may cause some qualitative differences from the small scale, smooth scenario studies by Tsuruya et al. and by Sato. The destructive power of the tsunami runup may however be estimated from the basically ballistic behaviour of the water near the front, i e, the uprush velocity through a level Z is of the order $\sqrt{2g(Z_{RL} - Z)}$ where Z_{RL} is the level of the runup limit. The runup limit is typically of the order 20m above MSL but extremes of 50–60m have been observed. Hence velocities of several metres per second can occur at elevations close to the runup limit and such velocities can destroy houses of light construction even if the local water depths are small.

Swell wave

breaking

on

Pearl Beach

2 Surf zone hydrodynamics

2.1 INTRODUCTION

The surf zone is the border area between the sea and the land within which the waves dissipate at least part of their energy through breaking. The surf zone is thus limited in the offshore direction by the point where the largest waves start to break. The location of this point will move with the tide and with changing wave parameters. At the land-ward end the surf zone is limited by the runup limit. That is, the most land-ward point reached by the wave runup.

Figure 2.1.1:
Depending on the wave parameters and the beach slope, the surf zone may be many wave lengths wide or very narrow.

Top:
The dissipative Durabah Beach at the entrance to the Tweed River.

Bottom:
The largely reflective Pearl Beach north of Sydney.

The surf zone may be divided into three sub areas:

(1) The outer surf zone where the waves change form rather rapidly after starting to break.

(2) The inner surf zone where all of the waves are breaking and an equilibrium breaker shape may exist if the bed slope is constant.

(3) The swash zone where the sand surface alternates between being exposed and being under water.

2.2 SURF ZONE WAVES

2.2.1 Breaking or reflection?
The extent to which the wave energy is dissipated by breaking as opposed to being reflected back offshore can be determined by the Iribarren Number or the surf similarity parameter ξ based on the deep water wave height

$$\xi_o = \tan \beta \sqrt{L_o / H_o} \qquad (2.2.1)$$

Waves will be predominantly reflected if $\xi_o > 4$. For smaller ξ_o most of the energy will be dissipated through breaking. See also Figure 1.9.4.

The rationale of this surf similarity parameter can be understood by considering the oscillation up and down a plane slope that corresponds to an incoming wave of height H and angular frequency ω.

Figure 2.2.1:
To deliver perfect reflection the water particles on the slope would need to oscillate along the slope with amplitude $\dfrac{H}{2}\dfrac{1}{\sin \beta}$.

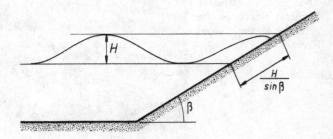

With angular frequency ω the maximum acceleration along the slope in this motion is of the order

$$A_{max} = \omega^2 \frac{H}{2} \frac{1}{\sin \beta} \qquad (2.2.2)$$

However, for the motion to be possible, this has to be less than the downslope acceleration that can be delivered by gravity:

$$A_g = g \sin\beta \qquad (2.2.3)$$

Otherwise the oscillatory motion is not possible and the wave must break. This leads to a criterion of the form $A_{max} < A_g$ or

$$\frac{g \sin \beta}{\omega^2 \dfrac{H/2}{\sin \beta}} \propto \frac{gT^2}{H} \sin^2 \beta \propto \frac{L_o}{H} \sin^2 \beta > \text{constant} \qquad (2.2.4)$$

for reflection rather than breaking.

Since the actual slope is $\tan\beta$ and $\sin\beta \approx \tan\beta$ for the small slopes that are typical for beaches the criterion is traditionally given in terms of the surf similarity parameter

$$\xi = \tan \beta \sqrt{L_o / H} \qquad (2.2.5)$$

In this expression H may alternatively be taken as the breaker height H_b or a deep water wave height H_o.

2.2.2 Breaker types

The surf similarity parameter also determines the breaking type. There are four main breaker types examples of which are listed in Table 2.2.2. The approximate ranges of $\xi_b = \tan \beta \sqrt{L_o / H_b}$ for which they occur on a straight slope are also listed.

Table 2.2.2.

Surf similarity parameter range	Breaker type
$4 < \xi_b$	Little or no breaking, cf Figures 1.9.4 and 7.2
$2 < \xi_b < 4$	Surging or collapsing breakers, see Figures 2.1.1 bottom and 6.10.1.
$0.4 < \xi_b < 2$	Plunging breakers, the surfable waves with transient shapes, see Figure 1.1.1
$\xi_b < 0.4$	Spilling breakers, may propagate with almost constant shape, Figure 2.1.1

2.2.3 Breaker heights

The limiting wave height to depth ratio for waves that propagate with constant form on a horizontal bed is theoretically limited to about 0.833, see Figure 1.5.6. However, waves that are approaching breaking on a slope may reach $H/h = \gamma$-values in excess of unity in the process. The maximum wave height and the maximum H/h may happen shortly before the wave actually breaks, i e, before part of the wave front becomes vertical and turns over.

Waves may also break in deep water. This occurs when the wave steepness is of the order 0.14 which is the limiting steepness for steady waves in deep water, see Figure 1.5.6.

A simple breaking criterion can be derived from $H_b = H_o K_r K_s = \gamma_b h_b$, which with the shallow water expressions for K_r and K_s leads to

$$H_b = \left[\frac{\gamma_b}{4k_o} \cos^2 \alpha_o H_o^4 \right]^{1/5} \qquad (2.2.6)$$

However, since γ_b depends on offshore wave conditions and beach profile shape this should only be taken as a rule of thumb. Svendsen (2006) gives a detailed discussion of wave breaking.

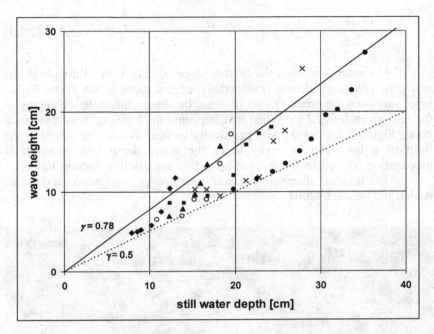

Figure 2.2.3: Measured heights of waves with different deep water parameters after breaking on a straight 1/65 slope. After Dally et al. (1985) based on data from Horikawa and Kuo (1966).

2.2.4 Wave height to water depth ratio after breaking

The wave height to depth ratio usually declines rather rapidly in the initial stages of breaking, see Figure 2.2.3. On a straight slope, an equilibrium with constant γ may be eventually be reached. The equilibrium γ-value tends towards a maximum of 0.55 for $\beta \to 0$.

In an equilibrium region, where $\gamma = H/h$ is constant (~ 0.5), the energy dissipation must relate in a special way to the beach slope. For example, assume that the energy flux in the breaking waves is analogous to that in a shallow water sine wave, cf Equation (1.4.41):

$$E_f = C_\beta \rho g H^2 \sqrt{gh} = C_\beta \rho g^{3/2} \gamma^2 h^{5/2} \qquad (2.2.7)$$

where C_β is a slope dependent constant of the order 1/8. Then, the rate of energy dissipation must be

$$D_E = -\frac{dE_f}{dx} = -\frac{5}{2}C_\beta\rho g^{3/2}\gamma^2 h^{3/2}\frac{dh}{dx} = \frac{5}{2}C_\beta\rho g^{3/2}\gamma^2 h^{3/2}\tan\beta \quad (2.2.8)$$

On natural beaches the bottom slope usually varies throughout the surf zone and may become considerably steeper close to the shore. In the inner surf zone the breakers may then not be able to dissipate the energy at the rate given by (2.2.8). When that happens, the breaking waves become bores, Figure 2.2.4 below, with practically vertical fronts. The height of the front of a bore can be many times the water depth into which it is propagating. The basic dynamics of bores was discussed in Section 1.5.3.

For a recent discussion of surf zone energy dissipation models see Alsina and Baldock (2007).

Figure 2.2.4: A surf zone bore carrying large amounts of suspended sand.

2.3 THE SHAPE OF THE MEAN WATER SURFACE

2.3.1 Terminology and definitions

The instantaneous water surface is traced by the instantaneous surface elevations $\eta(x,y,t)$ where x is shorenormal and y is shore parallel. The surface elevations are measured from the still water level (SWL) or rather, the still water surface (SWS), see Figure 2.3.1. The still water surface is the plane (on the scale considered), horizontal water surface that would exist in the absence of wind and waves. It moves up and down relative to the mean sea level (MSL) with the tide.

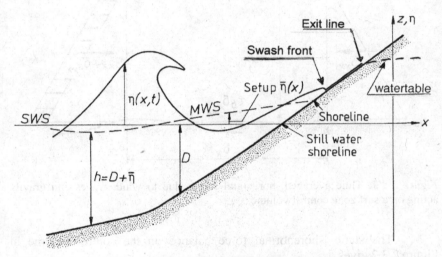

Figure 2.3.1: Definitions of levels and depths in the surf zone and the swash zone.

The still water depth D is measured from the local bed level to the SWS. The intersection between the SWS and the beach is the still water shoreline. The mean water level $\bar{\eta}(x,y)$ is the time average of $\eta(x,y,t)$ over a few minutes, i e, the corresponding mean water surface (MWS) does not move up and down with the wind waves but it does move with the tide. The total depth or just depth h is measured to the MWS.

As indicated by Figure 2.3.1, the MWS may be slightly below the SWS offshore due to setdown. In the surf zone it is above due to setup.

The exit line separates the glassy lower beach face from the upper matt looking part. It is the line along which the watertable meets the sand surface. The glassy looking area between the exit line and the swash front is the seepage face. The exit line may now and then be overrun by the swash fronts of larger waves.

The deviations of the MWS from the plane SWS are due to wind and waves and can be quantified on the basis of Figure 2.3.2.

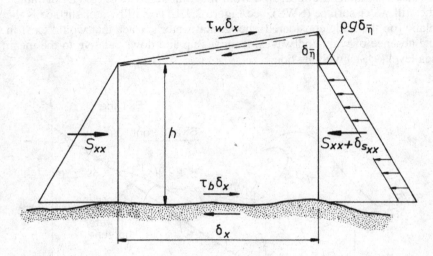

Figure 2.3.2: Time averaged, horizontal forces due to wind, waves and gravity acting on a surf zone control volume.

The static, shorenormal force balance on the control volume in Figure 2.3.2 gives

$$-\rho g \delta_{\bar{\eta}} h - \delta_{S_{xx}} + \tau_{\text{wind}} \delta_x - \tau_{\text{bed}} \delta_x = 0 \qquad (2.3.1)$$

or

$$\frac{\partial \bar{\eta}}{\partial x} = -\frac{1}{\rho g h} \frac{\partial S_{xx}}{\partial x} + \frac{1}{\rho g h}(\tau_{\text{wind}} - \tau_{\text{bed}}) \qquad (2.3.2)$$

2.3.2 Wave setdown

Outside the surf zone, the wave radiation stress will increase towards the shore as indicated by the sine wave explicit approximation

$$S_{xx} = \frac{3}{16}\rho gH^2[1-\frac{4}{9}k_oh+\frac{8}{135}(k_oh)^2] \qquad (2.3.3)$$

The trend emerges more clearly when this is rewritten in terms of the deep water wave height H_o by including the shoaling coefficient from linear wave theory

$$S_{xx} = \frac{3}{32}\rho gH_o^2\frac{1}{\sqrt{k_oh}}[1+\frac{1}{18}k_oh] \qquad (2.3.4)$$

This formula is within 1% of linear wave theory for $k_oh < 0.94$.

Inserting this expression for the radiation stress into the slope equation (2.3.2), neglecting the shear stresses, and integrating with the boundary condition $\bar{\eta} \to 0$ for $k_oh \to \infty$, we get the following expression

$$\bar{\eta} = -\frac{1}{32}k_oH_o^2(k_oh)^{-1.5}[1-\frac{1}{6}k_oh] \qquad (2.3.5)$$

for the setdown, i e, the lowering of the mean water level outside the surf zone. A setdown of this magnitude has often been measured in laboratory experiments with regular waves.

2.3.3 Wave setup

Inside the surf zone the radiation stress (proportional to H^2 for sine waves) will decrease rapidly towards the shore due to wave breaking and therefore generate a positive MWS slope through Equation (2.3.2).

A simple model of the wave setup in the surf zone can be obtained by ignoring the bed shear stress (its enhancing effect is analysed in Section 2.7) and adopting the shallow water sine wave relation between radiation stress and wave height

$$S_{xx} = \frac{3}{16}\rho gH^2 \qquad (2.3.6)$$

Neglecting the shear stress terms in (2.3.2) this leads to the slope

$$\frac{\partial\bar{\eta}}{\partial x} = -\frac{1}{h}\frac{3}{8}H\frac{\partial H}{\partial x} \qquad (2.3.7)$$

In order to solve this and get the setup as function of x or h, it is further necessary to assume a relation between wave height and depth. If we assume a constant ratio $H/h \equiv \gamma$, we obtain

$$\frac{\partial \bar{\eta}}{\partial x} = -\frac{3}{8}\gamma^2 \frac{\partial h}{\partial x} \qquad (2.3.8)$$

where we may insert $h = D + \bar{\eta}$ and get

$$\frac{\partial \bar{\eta}}{\partial x} = -\frac{3}{8}\gamma^2 \left(\frac{\partial D}{\partial x} + \frac{\partial \bar{\eta}}{\partial x}\right) \qquad (2.3.9)$$

and hence

$$\frac{\partial \bar{\eta}}{\partial x} = -\frac{\frac{3}{8}\gamma^2}{1+\frac{3}{8}\gamma^2}\frac{\partial D}{\partial x} = \frac{\frac{3}{8}\gamma^2}{1+\frac{3}{8}\gamma^2}\tan\beta \qquad (2.3.10)$$

That is, assuming linear wave theory and $\gamma \approx 0.5$ (cf Figure 2.2.3) for the breaking waves, we would expect a slope of the surf zone MWS of about $0.09\tan\beta$.

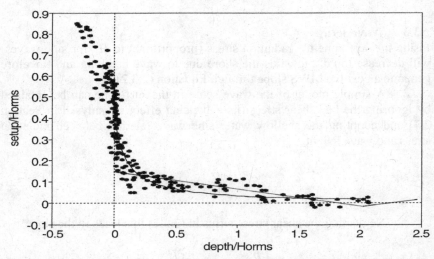

Figure 2.3.3: Setup profile in a natural surf zone. Measurements from South Beach, Brunswick Heads 22/6/1989.

Seeking the setup as function of the total depth, we can solve (2.3.8) with the boundary condition $\bar{\eta}(h_b) = \bar{\eta}_{\min}$ at the breakpoint and get

$$\bar{\eta}(h) = \bar{\eta}_{\min} + \frac{3}{8}\gamma^2(h_b - h) \qquad (2.3.11)$$

That is, according to the simple model based on regular waves, $S_{xx}(H)$ from linear wave theory and $H/h \equiv \gamma$, $\bar{\eta}$ varies linearly with h corresponding to the straight line in Figure 2.3.3.

The simple model leading to (2.3.11) is instructive, but it is well known to be an over simplification. The main discrepancy is that it predicts the setup to follow the decay of H^2 while data shows the setup to be somewhat "delayed". This delay was explained through the surface roller concept by Svendsen (1984). Svendsen suggested that S_{xx} remains undiminished for a while after wave breaking due to the momentum flux in a surface roller. The straight line in Figure 2.3.3 corresponds to Equation (2.3.11). That is, to regular waves, $S_{xx}(H)$ from linear wave theory and $H/h \equiv \gamma = 0.5$.

The deviation of the data from the straight line is mainly due to the fact that the real waves are irregular and start to break at different depths. Hence, the gradual increase in $\frac{d\bar{\eta}}{dh}$ compared to the regular wave model. The curve in Figure 2.3.3 corresponds to the empirical formula

$$\bar{\eta} = \frac{0.4 H_{\text{orms}}}{1 + 10\dfrac{h}{H_{\text{orms}}}} \qquad (2.3.12)$$

which provides a good fit to field data from a wide variety of beach morphologies.

2.3.4   Shoreline setup

The shoreline setup, i e, the setup at $h=0$ is predicted by the simple setup model (2.3.11) to be

$$\bar{\eta}_s = \bar{\eta}(0) = \bar{\eta}_{\min} + \frac{3}{8}\gamma^2 h_b = \bar{\eta}_{\min} + \frac{3}{8}\gamma H_b \qquad (2.3.13)$$

This will however, often be an underestimate of the actual shoreline setup under field conditions, see Figure 2.3.4.

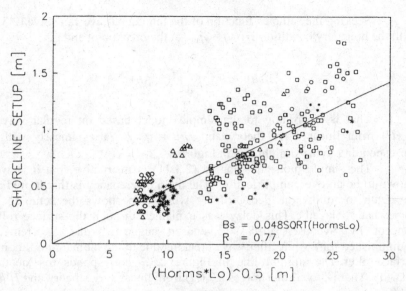

Figure 2.3.4: Observed shoreline setup on a wide variety of natural beaches. North Seven Mile Beach * , is among the flattest in New South Wales, Brunswick Beach □, is intermediate like Mid seven Mile, Dee Why Beach △, is similar to Narrabeen and Palm beach are usually very steep, like Collaroy, cf Figure 7.7.

The empirical curve (2.3.12) corresponds to

$$\bar{\eta}_s = 0.4 H_{orms} \tag{2.3.14}$$

and the same qualitative form, i e, $\bar{\eta}_s \propto H_{orms}$ results from the simple model (2.3.13). However, a slightly more reliable estimate is

$$\bar{\eta}_s = 0.048 \sqrt{H_{orms} L_o} \tag{2.3.15}$$

The straight line in Figure 2.3.4 corresponds to Equation (2.3.15). The data encompass almost the full spectrum of sandy beaches in New South Wales with beach face slopes in the range $0.03 < \tan\beta_f < 0.12$ during the experiments. There is no apparent dependence of $\bar{\eta}_s$ upon beach slope according to this data. However, based on a different data set, Stockdon et al. (2006) found improved correlation by including the beach slope. They recommended $\bar{\eta}_s = 0.35 \beta_F \sqrt{H_{orms} L_o}$.

2.3.5   Wave setup in river entrances

Flood modelling in coastal rivers starts with the tailwater level (the hydraulic control) at the river entrance. This obviously includes the astronomical tide but it may also include a considerable tidal anomaly caused by the weather system which generates the rainfall and hence the flood.

Professional engineers have traditionally (in the absence of detailed measurements) includeded some wave setup, usually a substantial fraction of the shoreline setup $\bar{\eta}_s$ discussed above in these tailwater levels. Usually the Bowen et al. model with its linear setup distribution, Equation (2.3.11), has been used to give the estimate $\bar{\eta}_s \times (h_b - h_{min})/h_b$, where h_{min} is the minimum depth and h_b the breaker depth. This however, is a serious overestimate. Partly because the real setup distribution is strongly upward concave, see Figure 2.3.2 but also because the wave crests are no longer straight between a pair of absorbing breakwaters, see Figure 2.3.5.

Figure 2.3.5:
Waves progressing between rock breakwaters at the Brunswick river entrance, NSW Australia. The wave crests become curved because of the energy absorption along the rock slopes.

The curvature of the wave crests and the corresponding flux of x-momentum in the y-direction, S_{xy}, adds an extra term to the balance equation (2.3.2) which then becomes

$$\frac{\partial \bar{\eta}}{\partial x} = -\frac{1}{\rho g h}\left(\frac{\partial S_{xx}}{\partial x} + \frac{\partial S_{xy}}{\partial y}\right) + \frac{1}{\rho g h}(\tau_{\text{wind}} - \tau_{\text{bed}}) \qquad (2.3.16)$$

Here, it turns out that the S_{xy}-term often almost cancels the S_{xx}-term. This effect of the wave dissipation via the crest curvature was investigated in detail by Dunn (2001), who also reviewed all the available data from river entrances on the East Coast of Australia, see Dunn et al. (2000).

Figure 2.3.6:
Scour hole on the beach side of the southern breakwater at Brunswick Heads. Large amounts of water and sand have been forced through the breakwater by the large waterlevel difference between the swash zone where $\bar{\eta} \approx \bar{\eta}_s$, and the river where $\bar{\eta} \approx 0$.

The experimental data however, shows even less contribution from wave setup than models, based on (2.3.16), predict. That is, for events with H_s up to 4m, the wave setup component in the Brunswick River has been unmeasurable (< 5cm) with the system described by Hanslow et al. (1992, 1996).

The scour hole shown in Figure 2.3.6 is also an indication of the lack of wave setup between the breakwaters.

2.3.6 Surf beat

As indicated by the wave records in Figure 1.6.1, natural waves tend to be groupy. The radiation stress $S_{xx} \sim H^2$ will be greater at the sets and may generate forced long waves in accordance with

$$\frac{\partial^2 \eta}{\partial t^2} = gh\frac{\partial^2 \eta}{\partial x^2} + \frac{1}{\rho}\frac{\partial^2 S_{xx}}{\partial x^2} \qquad (2.3.17)$$

as discussed in connection with Figure 1.3.4. In analogy with the solutions in Section 1.3, it follows that if the radiation stress associated with a set varies as $S_{xx}(x,t) = S_o f(x-c_g t)$, a steady forced wave given by

$$\eta_{forced}(x,t) = A_{forced}f(x-c_g t) = \frac{-S_o/\rho}{gh-c_g^2} f(x-c_g t) \qquad (2.3.18)$$

will be generated. If the forcing is switched on instantaneously, free long waves $\eta^+(x,t)$ and $\eta^-(x,t)$ with amplitudes

$$(A_{free}^+, A_{free}^-) = \left(-\frac{A_\infty}{2}[1+\frac{c}{\sqrt{gh}}], -\frac{A_\infty}{2}[1-\frac{c}{\sqrt{gh}}]\right) \qquad (1.3.33)$$

are generated as discussed in Section 1.3.3.4.

The forced wave expression (2.3.18) becomes singular in the shallow water limit where $c_g = \sqrt{gh}$. The linear problem then has no steady solution. Instead, $\eta^+(x,t)$ and $\eta_\infty(x,t)$ merge into a solution of the form $-t\frac{\partial}{\partial x}f(x-\sqrt{gh}\,t)$. Since $c_g = \sqrt{gh}\left[1-\frac{1}{2}k_o h+O(k_o h)^2\right]$, cf Table 1.4.16, the details are as follows

$$\eta(x,t) = \frac{-S_o/\rho}{c_g^2 - gh}\left(f(x - c_g t) - \frac{1}{2}[1 + \frac{c_g}{\sqrt{gh}}]f(x - \sqrt{gh}\,t) \right)$$

$$\xrightarrow[k_o h \to o]{} \frac{-S_o/\rho gh}{k_o h}\left(f(x - \sqrt{gh}[1 - \frac{k_o h}{2}]t) - f(x - \sqrt{gh}\,t) \right)$$

$$\xrightarrow[k_o h \to o]{} \frac{-S_o/\rho gh}{k_o h}\left(\sqrt{gh}\frac{k_o h}{2} t\, f'(x - \sqrt{gh}\,t) \right)$$

$$= \frac{-S_o}{2\rho\sqrt{gh}} t\, f'(x - \sqrt{gh}\,t) \qquad (2.3.19)$$

Hence, at resonant or near-resonant conditions, a Gaussian shaped forcing generates not a single depression in the form (2.3.18) but an N-shaped long wave. Because the forced wave amplitude (2.3.18) grows very strongly ($A_{\text{forced}} \sim h^{-5/2}$) as a set approaches the shore, the observed behaviour of the surf beat often resembles the resonant, abrupt onset solution (2.3.19) qualitatively.

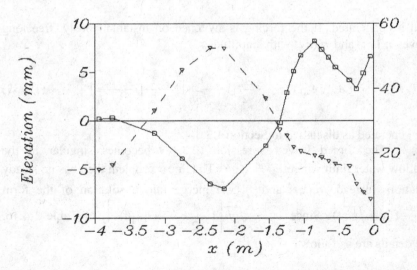

Figure 2.3.7: N-shaped long wave -□- and left hand scale together with the short-wave envelope -Δ- and right hand scale. Data from Baldock (2006), Beach slope 1/10, frequency range 0.8Hz–1.2Hz, H_o/L_o=0.088. x=0 is the still water shoreline.

Figure 2.3.7 shows an example of the long N-wave generated by a single short wave pulse in the laboratory. In the field things are always more complicated because of the continuous arrival of sets with different shapes and because of 2DH effects.

Figure 2.3.8: Backrush of strong surf beat. South Beach Brunswick Heads, 17/3/1992. Such events contribute a very large part of the beach erosion during storms.

2.4 SWASH AND RUNUP HEIGHTS

2.4.1 Introduction

The water motion in the area where the beach face is alternatingly exposed and covered by water on the time scale of individual wind waves is called swash, cf Figure 2.4.1.

If the surf similarity parameter (Equation (2.1.5)) is very large the waves will not break on the slope but for most natural beaches they will. The swash is then a highly asymmetrical motion where the depth and speed of the uprush is determined by the height and speed of the incoming bores while the backwash is driven by gravity. The sand, onto which the swash is progressing, may be saturated or unsaturated with water depending on the tidal phase and on the strengths of previous uprush events.

If the beach material is very coarse ($d_{50} > 1$mm), the swash may also be asymmetrical with respect to volume. That is, the amount of water running back down the slope is less than what rushed up because a substantial part of the uprush volume seeps into the slope.

Figure 2.4.1: Bores collapsing at the beach face and generating swash. In the foreground a bore which has not yet collapsed. Further back a smooth runup lens resulting from the collapse of a preceding bore.

On very flat beaches where breakers of constant form can dissipate energy fast enough to maintain constant H/h and hence $H{\to}0$ for $h{\to}0$, there will be no swash. The swash zone can be subdivided into an outer part where bores and swash lenses may overtake each other, and an inner part without interference, cf Hughes and Moseley (2007).

2.4.2   The hydrodynamics

The bore speed equation (1.5.9) blows up when the depth in front of the bore vanishes, i e, for $h_2{\to}0$.

What happens then is that a smooth swash lens shoots forward from the base of the bore front, see Figure 2.4.1. This has similarities with a dam break situation and the kinematics of the tip of the swash lens on a slope has been modelled successfully in analogy with dam breaks, using the non-linear shallow water equations. However, bore collapse differs from a dam break in that the water in the bore has initial momentum. This does not affect the kinematics of the tip, but it does affect the depths and flow rates following the tip as described by Guard and Baldock (2007).

The boundary layer structure and hence the shear stresses and the resulting sediment transport in the swash are still poorly understood. While some friction as opposed to ideal fluid flow makes a qualitative difference, the difference between a rough porous sand surface and a smooth plastic sheet makes little difference to the swash tip behaviour, see Figure 2.4.2.

Figure 2.4.2: Swash progressing onto a saturated sand surface, $d_{50}{\sim}0.5\text{mm}$, top and bottom, and onto a smooth plastic sheet between the stakes. The smooth, plastic surface has no obvious enhancing effect. South Palm Beach Sydney.

2.4.3 Runup heights for regular waves

When the incoming waves are regular, having fixed height and period, they will all run up to same height on the beach. In this case it makes sense to talk about the runup height R. It turns out that R depends in very different ways on the slope and the wave parameters in the breaking and non-breaking scenarios.

Experiments with breaking waves have shown that, the runup height measured from the still water level is proportional to the wave height H just seaward of the slope and to the surf similarity parameter

$$R = H\xi = H\tan\beta_F\sqrt{L_o / H} = \tan\beta_F\sqrt{HL_o} = \tan\beta_F\sqrt{H\frac{gT^2}{2\pi}} \quad (2.4.1)$$

In connection with this formula, we note, that the runup height depends more strongly on the wave period than on the height: $R \propto T\sqrt{H}$. Equation (2.4.1) which is often called Hunt's formula (Hunt 1959) is only valid for breaking waves with $\xi \le 2$ corresponding to $R \le 2H$, but this does cover the range of wind waves on natural beaches. For waves that do not break, the motion on the slope can be modelled with the non-linear shallow water equations. The result is stated by Meyer and Taylor (1972), based on the solution of Carrier and Greenspan (1958):

$$\frac{R}{H_o} = \sqrt{\frac{2\pi}{\tan\beta}} \quad \text{for } \xi > 4 \quad (2.4.2)$$

2.4.4 Runup height distribution for irregular waves

For natural waves with variable heights the maximum heights R_n on the slope reached by the individual swash fronts is variable.

Statistical information may be obtained by tracking the instantaneous waterline $\left(x_w(t), z_w(t)\right)$ with runup wires or a video camera. Runup wires are somewhat imprecise in that they have to be a certain distance above the sand surface, and even a video camera cannot always define a receding waterline when the surface remains wet due to seepage. Simple transgression statistics can be gathered by counting the number n_i of swash fronts passing stakes at sand levels z_i. These counts then define the distribution function for R_n, which is often expected to be a Rayleigh distribution:

126

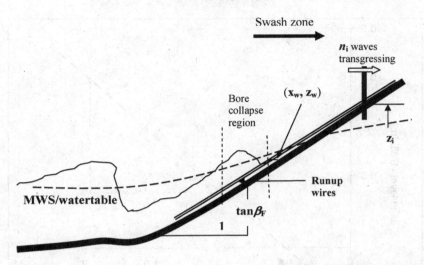

Figure 2.4.3: The region of bore collapse, has a finite width usually centred on the still water shoreline. The swash fronts may be monitored by runup wires or cameras. Alternatively, transgression statistics can be gathered as the numbers of swash fronts n_i passing stakes at different sand levels z_i, or by recording the heights to which lines in the sand get erased.

$$P\{R_n > x\} = \exp[-(\frac{x}{L_R})^2] \qquad (2.4.3)$$

This hypothesis corresponds to ($\sqrt{-\ln \frac{n_i}{N}}$, z_i) plotting along a straight line, where N is the total number of waves. The goodness of fit by the Rayleigh distribution, i e, a straight line in Figure 2.4.4 is variable. The slight upward convex trend near the runup limit in some of these datasets is perhaps a general feature. What is quite general, is that the best fit straight line does not go through the origin, i e, the point of 100% transgression does not generally coincide with the SWL.

The 100% transgression level varies quite widely, e g, $z_{100\%}/L_R = -0.31 \pm 0.40$ for the data of Nielsen and Hanslow (1991), see Figure 2.4.5.

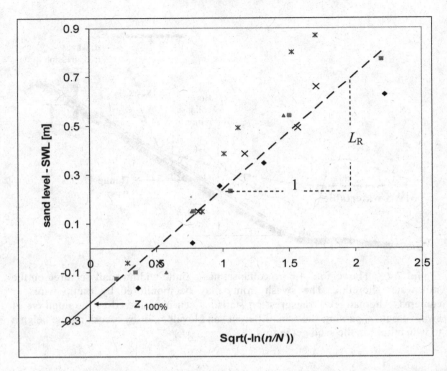

Figure 2.4.4: Rayleigh plots of five sets of runup data from Moreton Island, 10–11 December 2007. $H_{orms} = 0.9$m, $T_z = 5.6$s, $Tan\beta_F = 0.08$. The trend line corresponds to the data marked ■. $Z_{100\%}$ is the highest level transgressed by all waves according to the best fit Rayleigh distribution.

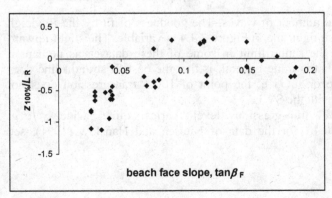

Figure 2.4.5: $z_{100\%}$ is close to zero ~SWL for $tan\beta_F > 0.05$, but it drops significantly on very flat beaches. Data from Nielsen and Hanslow (1991).

The very low values of $z_{100\%}$ observed on the flattest beaches are not due to swash zone processes. They correspond to bores merging in the inner surf zone on these beaches so that n_i starts to decrease well seaward of the average point of bore collapse.

The vertical runup scale L_R, which replaces H_{rms} from the Rayleigh wave height distribution in Equation (1.6.1), depends mainly on the offshore wave parameters and the beach face slope as indicated by Figure 2.4.6.

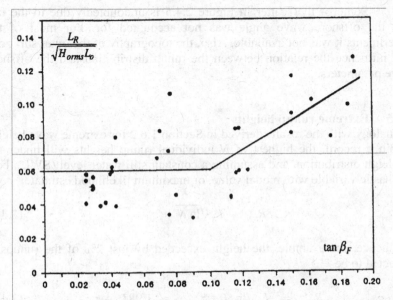

Figure 2.4.6: Relationship between the vertical scale L_R in the Rayleigh runup distribution and the offshore wave parameters. L_o is calculated from $_{sig} \approx T_z < T_p$. Data from Nielsen and Hanslow (1991).

The trend lines indicate that, for fairly steep slopes L_R is approximately 0.6 times the runup height of the rms wave calculated from regular-wave-equation (2.4.1):

$$L_R = 0.6 \tan\beta_F \sqrt{H_{orms} L_o} \quad \text{for } \tan\beta_F > 0.1 \qquad (2.4.4)$$

for flatter slopes L_R seems to be independent of β_F and approximately given by

$$L_R = 0.06\sqrt{H_{orms}L_o} \ \text{ for } \tan\beta_F < 0.1 \qquad (2.4.5)$$

However, this is an artifact of the transgression statistics being influenced by bores merging, and n_i correspondingly decreasing seaward of the bore collapse region on flat beaches. Correspondingly, Stockdon et al. (2006) found the 2% runup level to increase with beach slope throughout the range of slopes.

Some of the scatter in Figure 2.4.5 is undoubtedly due to the fact that the offshore wave angle was not accounted for. For most of the experiments it was not available. Also, the topography of the outer surf zone will influence the relation between the runup distribution and the offshore wave parameters.

2.4.5 Extreme runup heights
In analogy with the results derived in Section 1.6.2 for extreme wave heights within a record, the highest of N individual runup heights will under the Rayleigh distribution, and assuming a constant still water level (SWL), be a stochastic variable with modal value, or maximum likelihood estimate

$$R_{1/N} = L_R\sqrt{\ln N} + z_{100\%} \qquad (2.4.6)$$

and hence, for example, the height exceeded by just 2% of the runups is expected to be

$$R_{2\%} = L_R\sqrt{\ln 50} + z_{100\%} \approx 1.98L_R + z_{100\%} \qquad (2.4.7)$$

The question of extreme runup height is more complicated if the SWL is varying due to tide or storm surge or a combination of those. The expected height of the runup limit, will then depend on how rapidly the SWL varies around its peak. Consider the scenario in Figure 2.4.7, where a surge+tide with arbitrary but fairly smooth shape is superposed by wave runup.

We wish to estimate height of the runup limit RL. This can of course be done for an arbitrary surge shape with Monte Carlo simulation. However, since RL will depend mainly on the maximum surge level η_{max} and the radius of curvature of the surge near the peak, we will develop a more direct practical estimate. This estimate is based on a Rayleigh runup process with

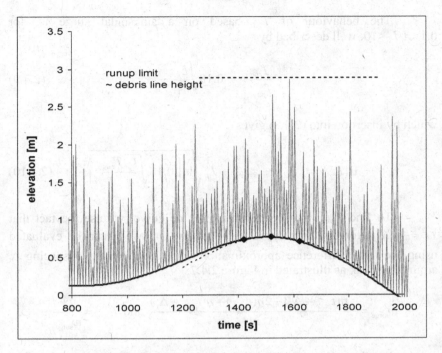

Figure 2.4.7: The smooth solid curve shows storm surge + tide, on which a Rayleigh runup process with vertical scale L_R=1m, $z_{100\%}$=0 and wave period T=10s is superimposed. The time scale of the SWL variation around the peak can be defined as the period of a sine curve (thin dotted curve), which approximates the shape of the surge+tide around the peak.

wave period T and vertical scale L_R together with a sinusoidal surge with period T_{surge} and amplitude A. We aim for an estimate in the form

$$E\{RL\} = \eta_{max} + z_{100\%} + L_R\sqrt{\ln\frac{T_{eff}}{T}} \qquad (2.4.8)$$

The rationale for the last term is that the expected runup limit for N waves in a steady Rayleigh process (no tide or surge) with vertical scale L_R is $E\{RL\} = L_R\sqrt{\ln N}$ and, T_{eff}/T is the number of waves in an effective time span around the surge peak.

The behaviour of T_{eff} based on a sinusoidal surge is, for $0.1 < A/L_R < 10$, well described by

$$\frac{T_{eff}}{T_{surge}} = 0.18\sqrt{\frac{L_R}{A}} \tag{2.4.9}$$

which by insertion into (2.4.8) gives

$$E\{RL\} = \eta_{max} + z_{100\%} + L_R\sqrt{\ln\left(0.18\sqrt{\frac{L_r}{A}}\frac{T_{surge}}{T}\right)} \tag{2.4.10}$$

To find T_{surge} by approximating a sine curve, we use the fact that $\omega^2 = -y''/y$ for any sine curve $y = A\sin(\omega t - \varphi)$, which can be evaluated using the finite difference approximation with three points, of spacing Δ_t around the peak as illustrated in Figure 2.4.7:

$$\omega^2_{surge} = -\frac{\eta(t_{peak} + \Delta_t) - 2\eta(t_{peak}) + \eta(t_{peak} - \Delta_t)}{\Delta^2_t \, \eta(t_{peak})} \text{ and } T_{surge} = \frac{2\pi}{\omega_{surge}}.$$

For the example in Figure 2.4.7, this procedure gives $T_{surge} = 1856$s, and with $A = \eta_{peak} = 0.768$m, $L_R = 1$m, $z_{100\%} = 0$ and $T = 10$s Equation (2.4.10) gives $E\{RL\} = 2.68$m.

Perhaps because of the sometimes poor fit of the Rayleigh distribution and the uncertainty about $z_{100\%}$, Stockdon et al. (2006) took the 2% runup level as their estimate of the extreme runup level. This is perhaps primitive statistics, which is not readily adaptable to runup on top of a time varying surge level, but it has the advantage of being free of assumptions about the type of distribution function.

2.5 WAVE GENERATED LONGSHORE CURRENTS

2.5.1 Introduction

When the waves break at an angle with the coast, they often generate shore parallel currents in the surf zone. These wave driven longshore currents may coexist with other shore parallel currents driven by wind or tide.

The source of momentum, or the driving force, for the wave driven longshore currents is the influx of shore parallel wave momentum flux at the breaker line, see Figure 2.5.1. In the steady situation this driving force is balanced by the shore parallel, time averaged bed shear stress $\overline{\tau}_{by}$.

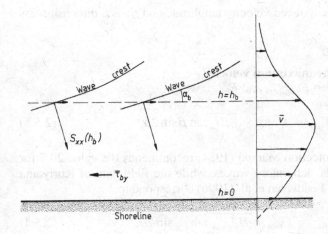

Figure 2.5.1:
Longshore currents are driven by the shore parallel component of the momentum flux in the breaking waves.

2.5.2 The driving force

The wave momentum flux per unit length of the breaker line is $F_m(h_b)\cos\alpha_b$ [N/m] and the longshore component is

$$F_{\text{longshore}} = F_m(h_b)\cos\alpha_b\sin\alpha_b \propto \rho g H_b^2 \sin 2\alpha_b \qquad (2.5.1)$$

cf Equation (1.4.52). The pressure component of the radiation stress is isotropic, i e, it has no direction, and therefore it makes no contribution to the net force in the y direction.

In a steady situation, this driving force will be balanced by the total bed friction force per unit length of surf zone.

2.5.3 The retarding force

In terms of the maximum longshore velocity V_{max}, and the surf zone width $h_b/\tan\beta$ the total friction force can reasonably be assumed to have the form

$$F_{friction} = -\rho f_{cw} \left| u_{max} \right| V_{max} \frac{h_b}{\tan\beta}$$

$$(2.5.2)$$

$$\propto -\rho\sqrt{gH_b} \, V_{max} \frac{H_b}{\tan\beta}$$

where u_{max} is the wave induced velocity amplitude and f_{cw} is a dimensionless friction factor.

2.5.4 Estimating the maximum velocity

From a balance between $F_{longshore}$ and $F_{friction}$ we get

$$V_{max} = \text{Const} \times \sqrt{gH_b} \, \tan\beta \sin 2\alpha_b \qquad (2.5.3)$$

The Shore Protection Manual (1984) recommends the value 20.7 for the constant for regular laboratory waves, while the field data of Kuriyama and Ozaki (1993) and Feddersen et al. (1996) correspond to

$$V_{max} = C_{field}\sqrt{gH_{sig,max}} \, \tan\beta_S \sin 2\alpha_b \qquad (2.5.4)$$

with

$$C_{field} \in [12;30] \qquad (2.5.5)$$

where $H_{sig,max}$ is the peak significant wave height at the edge of the surf zone and $\tan\beta_S$ is the average bed slope in the surf zone, calculated as the depth where the wave height peaks divided by the distance from this point to the shoreline.

2.5.5 Distribution of the longshore current velocity

The distribution of the longshore current in a simple surf zone with constant bed slope, linear shallow water waves and constant H/h has been instructively discussed by Longuet-Higgins (1972). However, the distribution in real surf zones is still not fully understood.

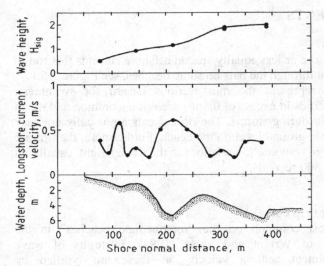

Figure 2.5.2: Measured wave heights and time-mean longshore current velocities. The strongest currents do not coincide with the steepest wave height gradient. After Kuriyama and Ozaki (1993).

While Longuet-Higgins' (and most others') theory predicts maximum current speed where the wave height gradient, and hence the forcing, which is assumed to be proportional to $\dfrac{\partial H^2}{\partial x}$, is greatest, field data tend to show maximum current over the deepest part of the trough, see Figure 2.5.2.

The challenges lie in predicting the actual variation of the wave momentum flux across the surf zone and the nature of the momentum transfer between vertical, shore parallel slices of water.

Some indication about the momentum flux distribution can be gained from observed surf zone setup profiles (cf Section 2.3.3 and Hanslow et al. 1996). The "delay" of the current peak compared with Max{dH/dx} is related to the similar delay in the setup gradient, cf p. 61. It is believed to be due to the surface roller contributing considerably to the momentum fluxes as described by Svendsen (1984). See also Feddersen et al. (1996, 2004).

Some recent theoretical advances on the redistribution of y-momentum between shore parallel slices of water, have been made by Svendsen and Putrevu (1994), Svendsen (2006).

The wave driven longshore currents are often unsteady, varying on the time scale of wave groups, i e, 100 to 300 seconds.

2.6   RIP CURRENTS

2.6.1   Introduction

Rip currents are the more or less equally spaced offshore currents that run in the channels which cut through the bars on some beaches, see Figure 2.6.1.

Rip currents represent the most serious hazard for swimmers. Seaward current velocities in excess of 0.5m/s are not uncommon and much stronger currents occur during storms. The rip currents are usually not easy to see for an observer at ground level on the beach. Furthermore, the rips are not steady but pulsate in response to variations in the wave height, usually on a time scale of 100 to 300 seconds.

2.6.2   Occurrence of rip currents

Rip currents are most common on the "intermediate beaches" in the classification system of Wright and Short (1984). In terms of wave parameters and sediment settling velocity w_s these are typified by $1 < \dfrac{H_b}{w_s T} < 6$, cf Section 7.2.

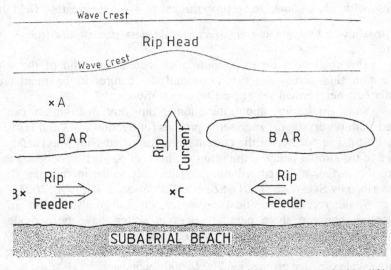

Figure 2.6.1: The rip system is driven by waves pumping water across the bar, lifting the waterlevel at B above the offshore waterlevel. This water then flows along the rip feeder towards C and out through the rip.

136

2.6.3   Rip spacing

The usual spacing of rip currents on a long straight coast is of the order two to five times the distance from the shore to the breakpoint. Hence, in general, larger waves and/or flatter beach profiles mean larger rip spacing. Consequently, an embayed beach, less than a kilometre long may have several rip cells when the waves are small but only a single "mega rip" during storms.

The position of rip currents is often topographically forced by pre-existing bars or harder constraints like headlands, groins and breakwaters. There are also many examples of permanent rips under piers supported by piles.

2.6.4   Rip flows and wave pump efficiency

The driving mechanism for the rip systems is that the waves push water across the bar generating an average MWS over-height $\Delta_{\bar{\eta}}$ inside the bar. If the flow rate for the system is Q [m³/s], this corresponds to the useful power $\rho g \Delta_{\bar{\eta}} Q$ delivered by the waves as a pump. The efficiency of the wave pump, which drives the rip is then defined as the useful power divided by the incoming wave power:

$$E_{\text{rip}} = \frac{\rho g \Delta_{\bar{\eta}} Q}{\frac{1}{8} \rho g H^2 \sqrt{gh_b}\, L_{\text{rip}}} \tag{2.6.1}$$

where L_{rip} is the alongshore length of the rip system.

Values of E_{rip} from field and laboratory experiments vary surprisingly little despite considerable difference in topography. Thus, Nielsen et al. (2001) found $E_{\text{rip}} \in [0.03;0.05]$ from field data with $H_{\text{orms}}<1.5$m and laboratory experiments where the "bars" were sharp crested concrete ramps.

Figure 2.6.2: Two large rip systems at Tallow Beach south of Cape Byron, Australia made visible by the light coloured rip heads outside the surf zone. The rip currents themselves are not clearly visible even from this elevated position. Standing on the beach or in the water it is impossible to see where the rip currents are.

2.7   UNDERTOW

The water level near the shore will generally be higher than at the breaker line due to wave setup as discussed in Section 2.3.3. This generally leads to a 2DV circulation as shown in Figure 2.7.1: The flow will be on-shore near the surface where the wave induced, time averaged momentum flux $\overline{\rho u^2}$ is concentrated. Near the bed, the pressure gradient generated by the MWS slope will however dominate and drive a seaward current. This seaward current below the wave trough level is often referred to as undertow.

Both undertow and rip currents are thus driven by $-dS_{xx}/dx$ pushing water onshore. In a longshore uniform scenario the water escapes along the bottom as undertow, while it usually gets concentrated in rip channels on natural beaches.

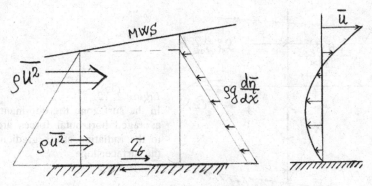

Figure 2.7.1: The concentration of the wave induced momentum flux near the surface combined with the vertically uniform seaward pressure gradient generate shoreward net flow near the surface and seaward net flow called undertow near the bed.

Associated with the undertow is a seaward shear stress $\overline{\tau}_b$ on the bed, which increases the shoreline setup through the general force balance (Figure 2.3.2) and generates offshore sediment transport. The magnitude of $\overline{\tau}_b$ is of the order $(1/4)\mathrm{d}S_{xx}/\mathrm{d}x$ depending on the vertical distributions of $\mathrm{d}S_{xx}/\mathrm{d}x$ and the eddy viscosity as illustrated by the example below.

Example 2.7.1:
Assume that the radiation stress S_{xx} and hence its gradient are linearly distributed between the bed and the mean water surface so that the force from $\mathrm{d}S_{xx}/\mathrm{d}x$ corresponds to the triangle in Figure 2.7.2, while the facing rectangle represents the pressure force due to the MWS slope $\dfrac{\mathrm{d}\overline{\eta}}{\mathrm{d}x}$. The resultant force above a level z then gives the shear stress:

$$\overline{\tau}(z) \neq -\frac{\mathrm{d}S_{xx}}{\mathrm{d}x}[1-(\frac{z}{h})^2]-\rho g\frac{\mathrm{d}\overline{\eta}}{\mathrm{d}x}h[1-\frac{z}{h}] \qquad (2.7.1)$$

from which the velocity distribution can be found by integration:

$$\overline{u}(z) = \int_{z=0}^{z} \frac{\overline{\tau}(z)}{\rho v_t}dz .$$

139

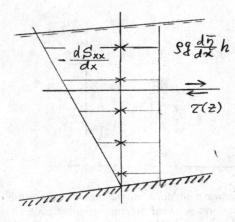

Figure 2.7.2:
In the surf zone the dominant time averaged horizontal forces are due to the radiation stress gradient and the surface slope.

With a constant eddy viscosity gives

$$\bar{u}(z) = \frac{1}{\rho v_t}\left[-\frac{dS_{xx}}{dx}[z - \frac{1}{3h^2}z^3] - \rho g\frac{d\bar{\eta}}{dx}[hz - \frac{1}{2}z^2]\right] \qquad (2.7.2)$$

For a specified flow rate this determines the MWS slope. In particular, for the longshore uniform situation where $\int_{z=0}^{h} \bar{u}\,dz = 0$, one finds

$$\frac{d\bar{\eta}}{dx} = -\frac{5}{4\rho gh}\frac{dS_{xx}}{dx} \qquad (2.7.3)$$

i e, 5/4 times the result (2.3.2) which was obtained without consideration of the undertow and the associated shear stresses. Correspondingly, the undertow-related seaward bed shear stress is

$$\bar{\tau}_b = \frac{1}{4}\frac{dS_{xx}}{dx} \qquad (2.7.4)$$

for the case of $\int_{z=0}^{h} \bar{u}\,dz = 0$ and constant v_t.

Note that (2.7.4) is independent of the magnitude of v_t.

☺

140

3 Wind driven circulation and storm surges

3.1   INTRODUCTION

Coastal storm hazards may be simple in the sense that only one destructive agent is at play, e g, tsunami which were described in Section 1.11. More commonly however, coastal hazards result from a combination of oceanic surges on the scale of the whole weather system, i e, hundreds of kilometres, and wind wave effects which are generated between the break point and the runup limit, i e, within a few hundreds of metres, see Figure 3.1.1.

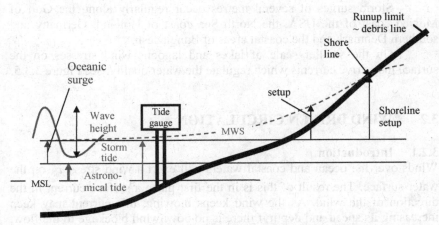

Figure 3.1.1: Storm surge terminology. The oceanic surge is uniform over the surf-zone. The MWS variation in the surf zone is thus due to waves and onshore winds.

On the scale of the surf zone, the oceanic surge can be considered uniform. However, wave radiation stress and local wind stress will cause MWS variations, setup, inside the surf zone. Tide gauges are usually in depths of several metres where the latter effects are insignificant.

While a lot of damage to coastal areas is caused by winds, we shall concentrate on the damage caused by coastal waters: High storm tide levels may disrupt road transport, and in the longer term cause salt damage to agricultural land. Houses are however usually destroyed by wind waves either through runup impact, Figure 1.6.13, or through undercutting erosion, Figure 7.1.3.

As described in Section 1.3.3, the pressure variations and wind stresses due to weather systems can generate long waves, on the scale of the weather systems themselves. These waves will, because they are very long, often be influenced by the Earth's rotation through the Coriolis effect, as well as by gravity. The geostrophic effect, i e, the influence of the Earth's rotation, is most strikingly seen by the fact it only allows anticlockwise travel around continents of storm surges in the southern hemisphere (clockwise in the northern hemisphere). Thus, waves generated by cyclones on the North West Shelf of Australia often travel anticlockwise around the coast to Bass Strait, see Figure 1.3.3, while cyclones off the coast of Queensland never send long waves southward. Long waves may analogously be trapped along the inside of "bowls" like the North Sea. Here storm surges are often observed to travel anticlockwise, down the coast of Great Britain, across to Holland and up the west coast of Denmark.

Storm surges of several metres occur regularly along the Gulf of Mexico coast of the USA, the North Sea coast of Holland, Germany and southern Denmark and the coastal areas of Bangladesh.

On the smaller scale of lakes and lagoons wind stresses on the surface may drive currents which regulate the water quality, see Figure 5.3.5.

3.2   WIND DRIVEN CIRCULATION

3.2.1   Introduction

Winds over the ocean and coastal waters will exert a wind stress τ_w on the water surface. The result of this is in the first place a surface current in the direction of the wind. As the wind keeps blowing, this current may keep increasing its speed and depth if there is no downwind blockage to the flow, see Section 3.2.5 for Coriolis effects in very deep water.

If the wind blows towards the land so that the flow is blocked at the shore, the nearshore water level will rise and the corresponding pressure gradient will drive a return current near the bottom, which eventually balances the onshore flow near the surface, see Figure 3.2.1.

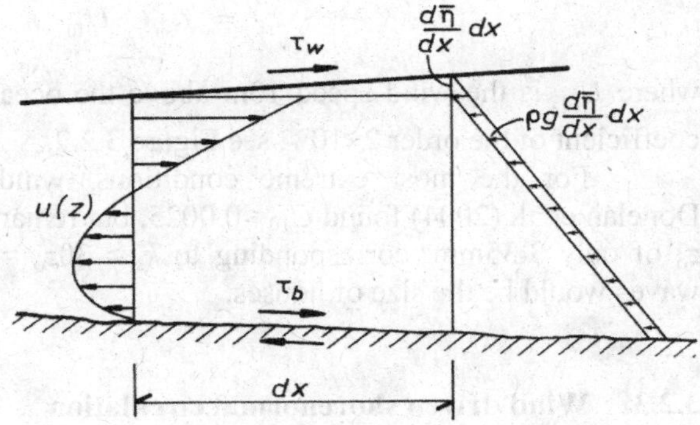

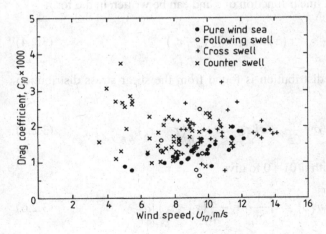

Figure 3.2.1:
Wind stress, wind setup and wind driven circulation.

If the wind is gusting, both currents and surge levels may oscillate and if waves at the gusting frequency resonate with the topography, see Section 1.9.2 on seiches, very strong oscillations may occur.

3.2.2 Wind stresses on the sea surface

The wind stress τ_w on the ocean surface depends on the size and shape of the waves as well as on the wind speed. Hence, τ_w will be different offshore and in the surf zone on a given day.

Not much is known about wind stresses on surf zones, but for near-equilibrium waves in deeper water τ_w can be estimated from

Figure 3.2.2:
Wind drag coefficients measured under different wave conditions in deep water — not surf zones. After Donelan et al. (1997). For recent reviews, see Csanady (2001) and Jones and Toba (2001).

143

$$\tau_w = \rho_{\text{air}} C_{10} U_{10}^2 \qquad (3.2.1)$$

where U_{10} is the wind speed 10m above the ocean surface and C_{10} a drag coefficient of the order 2×10^{-3}, see Figure 3.2.2.

For the most extreme conditions, wind speeds beyond 33m/s, Donelan et al. (2004) found $C_{10} = 0.0025$, but remarkably, a roughness length z_0 of only 3.35mm, corresponding to $k_s \approx 30 z_0 \approx 0.1$m. This is while the waves would be the size of houses.

3.2.3 Wind driven shorenormal circulation

Under the assumption of steady 2DV conditions, the wind setup and the corresponding 2DV circulation pattern can be analysed in the terms of Figure 3.2.1. That is, a steady force balance on the control volume requires

$$\rho g \frac{d\overline{\eta}}{dx} h = \tau_w - \tau_b \qquad (3.2.2)$$

In a steady situation we always have

$$\frac{\partial p}{\partial x} = \frac{\partial \tau}{\partial z} \qquad (3.2.3)$$

where the pressure gradient is constant over the depth if the pressure is hydrostatic.

Hence, τ is a linear function of z and can be written in the form

$$\tau(z) = \tau_b + \frac{z}{h}(\tau_w - \tau_b) \qquad (3.2.4)$$

The velocity distribution is found from the shear stress distribution through

$$\rho v_t \frac{\partial u}{\partial z} = \tau(z) = \tau_b + \frac{z}{h}(\tau_w - \tau_b) \qquad (3.2.5)$$

which is integrated with $u(0) = 0$ to give

$$u(z) = \int_0^z \frac{1}{\rho v_t} \left[\tau_b + \frac{z}{h}(\tau_w - \tau_b) \right] dz \qquad (3.2.6)$$

In steady, strictly two-dimensional scenarios there must be zero net flow in a vertical: $\int_{bed}^{surface} u\,dz = 0$, which leads to a scenario like the one in Figure 3.2.1.

3.2.4   Wind driven steady circulation in a basin

In most natural scenarios, the flow is of course three-dimensional and the Coriolis effect becomes important on larger scales. An example of moderate scale 3D wind driven circulation is shown in Figure 3.2.3. The numbers on the circular depth contours are in metres.

Note that the depth integrated flow is down-wind in the shallow sections but essentially upwind in the deeper sections. The directional deviations from the wind direction are due to the Coriolis effect.

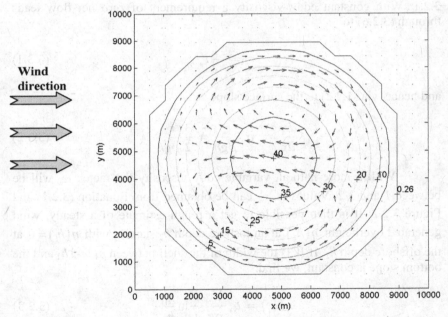

Figure 3.2.3: Wind driven flow rates (vertically integrated velocities) in a circular lagoon with parabolic depth. Calculations by Dr Fred Saint Cast, $\tau_w = 0.048 \text{N/m}^2$, $v_t = 0.092(h/h_{max})$ [m²/s], latitude 14.5° South.

145

3.2.5   Steady wind driven circulation in deep water

If the wind is steady for several days and the water is deep, the Coriolis effect generates the Ekman spiral, Ekman (1905). For constant eddy viscosity v_t that is: The surface current is $45°$ clockwise from the surface wind stress in the northern hemisphere, opposite in the southern hemisphere, and the angle increases downwards while the current magnitude decreases exponentially on the vertical scale $\sqrt{\Omega v_t}$, where Ω is the Earth's angular frequency.

3.3   SETUP AND SURGES DUE TO WIND STRESS FIELDS

3.3.1   Wind setup from a steady onshore wind

The following is a brief, quantitative description of elevated coastal water levels due to steady onshore winds. The scenario is again that of Figure 3.2.1. With constant eddy viscosity a requirement of zero net flow leads through (3.2.6) to

$$\tau_b = -\frac{1}{2}\tau_w \qquad (3.3.1)$$

and hence, via (3.2.2) to the surface slope

$$\frac{d\overline{\eta}}{dx} = \alpha_v \frac{\tau_w}{\rho g h} = \frac{3}{2}\frac{\tau_w}{\rho g h} \qquad (3.3.2)$$

With a more natural, variable eddy viscosity the factor α_v will be between 1 and 3/2. Values of τ_w can be obtained from Equation (3.2.1) and Figure 3.2.2. It is then possible to get a rough estimate of a steady, wind generated overheight $\overline{\eta}(h_2)$ at an inshore depth h_2 starting with $\overline{\eta}(h_1) = 0$ at the offshore depth h_1. If W is the width of the shelf between h_1 and h_2 and the bottom slope is constant, we find:

$$\overline{\eta}(h_2) = \alpha_v \frac{\tau_w}{\rho g}\frac{W}{h_1}\ln\frac{h_1}{h_2} \qquad (3.3.3)$$

This formula has, unfortunately, no meaning for $h_2 \rightarrow 0$, hence the nearshore setup has to be evaluated at some arbitrary finite depth. Another problem with it is that the drag coefficients for surf zones may very well be

146

greater than the deep water values shown in Figure 3.2. Nevertheless, the formula shows the important fact that "the wider the shelf the greater the wind setup".

3.3.2 Surge due to steady longshore wind

If the continental shelf is steep the effect of shorenormal winds is, at most, only a few centimetres as indicated by (3.3.3). However, the Coriolis or geostrophic effect on large scale wind driven longshore currents may in this case cause significant tidal anomalies. The Coriolis acceleration on a water particle with horizontal velocity u is perpendicular to u, to the left in the southern hemisphere and to the right in the northern hemisphere, cf Alonso and Finn (1968) or LeMehaute (1976), and its magnitude is

$$a_{\text{Coriolis}} \quad = \quad 2\omega_E u \sin\theta \qquad (3.3.4)$$

where $\omega_E = 7.3 \times 10^{-5}$ rad/s is the angular frequency of the Earth's rotation and θ is the latitude. The typical magnitude of the Coriolis acceleration is, according to (3.3.4), 10^{-4} m/s^2 or $10^{-5}g$ meaning that, it can (with current speeds of the order 1m/s) induce ocean surface slopes of the order 10^{-5} or 1cm per kilometre. This often becomes significant on the scale of weather systems (>100km).

3.3.3 Surge due to a moving wind stress field

The scaling and basic dynamics of surges due to moving wind stress fields is illustrated by a simple 1D analytical example, based on the general solution scheme for constant depth in Section 1.3. Subsequently differences and similarities with 2D scenarios are illustrated and discussed briefly.

Consider the wind stress field

$$\tau_w(x,t) = \frac{\tau_{\max}}{\cosh^2(\frac{x-ct}{L})} \qquad (3.3.5)$$

which in the notation of Section 1.3.3.2 corresponds to "the stress potential"

$$T_w(x,t) = \tau_{\max} L \tanh(\frac{x-ct}{L}) \qquad (3.3.6)$$

i e, we have $\tau_w = \dfrac{\partial T_w}{\partial x}$. And we found that the steady forced wave solution is then through (1.3.23) given by

$$\eta_{\text{forced}}(x,t) = A_{\text{forced}} \tanh(\frac{x-ct}{L}) = \frac{\tau_{\text{max}}L}{\rho(gh-c^2)} \tanh(\frac{x-ct}{L}) \qquad (3.3.7)$$

and the complete solution, including free waves, for a surge which grows after the onset of the forcing at $t = 0$:

$$\eta(x,t) = \eta_{\text{forced}} + \eta_{\text{free}}^+ + \eta_{\text{free}}^-$$

$$= A_{\text{forced}} \tanh(\frac{x-ct}{L}) + A_{\text{free}}^+ \tanh(\frac{x-\sqrt{gh}\,t}{L}) + A_{\text{free}}^- \tanh(\frac{x+\sqrt{gh}\,t}{L})$$

$$(3.3.8)$$

which with (3.3.6) and (1.3.23) can be rewritten as

$$\eta(x,t) = \frac{\tau_{\text{max}}L}{\rho(gh-c^2)} \left(\begin{array}{c} \tanh(\dfrac{x-ct}{L}) - \dfrac{1}{2}[1+\dfrac{c}{\sqrt{gh}}]\tanh(\dfrac{x-\sqrt{gh}\,t}{L}) \\[4mm] -\dfrac{1}{2}[1-\dfrac{c}{\sqrt{gh}}]\tanh(\dfrac{x+\sqrt{gh}\,t}{L}) \end{array} \right) \qquad (3.3.9)$$

An example is shown in Figure 3.3.1. For these wind stress driven surges, the asymptotic state does not correspond to the forced wave because the free waves have finite surface elevations at $x \to \pm\infty$.

We note that the general result from Section 1.3.3.5 that the resonant 1D surge takes the shape of $t\sqrt{gh}\dfrac{\partial}{\partial x}f(x-ct)$ where $f(x-ct)$ is the shape of the steady, non-resonant forced solution means that at resonance, a wind stress driven 1D surge takes the form of $t\tau_w$ rather than of T_w.

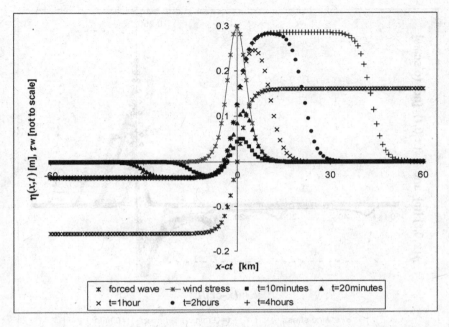

Figure 3.3.1: Developing surge driven by a pulse shaped wind stress travelling at speed c =11m/s. τ_{max}=3Pa, L=4000m, h=20m, $\sqrt{gh}$ =14m/s. A_{forced}=0.160m, A_{free}^{+} = −0.143m, A_{free}^{-} = −0.017m.

The asymptotic behaviour of 2D wind stress driven surges is different from that of the 1D solution in Figure 3.3.1. That is, the plateaus in front of and behind the stress field are not flat but have elevated edges, see Figure 3.3.2.

In addition we note that the 2D surge is generally smaller because of radiation in the y-direction. It is also interesting that the 2D solution overshoots at the centre compared with the asymptotic central value.

The fact that a wind stress of a given magnitude generates bigger surges in shallower water, cf Equation (3.3.7), means that the landfall scenario of fast moving surges typically corresponds to the early stages in Figure 3.3.2, e g 19min ~ ■.

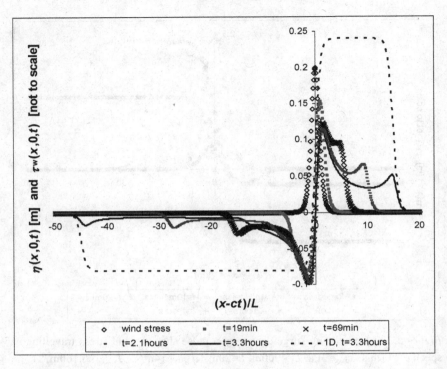

Figure 3.3.2: Sections along $y=0$ through a developing 2D wind stress surge of the

form $\quad \tau_w(x,y,t) = \dfrac{\tau_{max}}{[\cosh(\sqrt{[x-ct]^2 + y^2}\,/\,L)]^2}\quad$ with $\quad \tau_{max}=3\text{Pa}, \quad h=10.2\text{m},$

$\sqrt{gh} = 10\text{m/s}$, $c= 5\text{m/s}$ and $L= 4000\text{m}$, compared with the 1D solution (3.3.8).

3.4 BAROMETRIC SETUP AND PRESSURE DRIVEN SURGES

3.4.1 The stationary scenario

Basic hydrostatic considerations can be used to show that a stationary low pressure system will raise the sealevel in a matching shape and with local over-height given by $\Delta_\eta = -\Delta_p/\rho g$, irrespective of the water depth. This corresponds to the rule of thumb that: For every hPa of low pressure one gets 1cm of local sealevel rise.

3.4.2   Long waves forced by a moving low pressure system

Interestingly, if the low pressure system moves, one can get considerably more storm surge than the abovementioned 1cm/hPa. This can be understood qualitatively in terms of the theory for forced long waves with small amplitude which was developed in Section 1.3.3. We found there that a low pressure system which propagates with constant but arbitrary form, e g, $p_s(x,t) = P_o f(x-ct)$ will generate a steady forced wave of the form $\eta_{\text{forced}}(x,t) = A_{\text{forced}} f(x-ct)$, where

$$A_{\text{forced}} = \frac{-P_o/\rho g}{1-c^2/gh} \qquad (1.3.19)$$

See Figure 3.4.1.

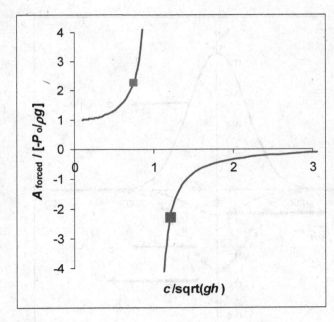

Figure 3.4.1: Amplitude of a steady surge moving with speed c compared to that of a stationary surge according to Equation (1.3.19). For a slow low pressure system the steady response is a positive surge. For a fast system it is a negative surge.

That is, a low pressure system ($P_o<0$) which moves slowly: $c < \sqrt{gh}$, generates a positive surge, while a fast moving low pressure system generates a negative surge in the steady state. The latter does however not mean that fast systems cannot generate flooding. The flooding is in that case generated by the forward moving free wave $\eta^+(x,t)$, which is

151

borne together with $\eta_{\text{forced}}(x,t)$ at the onset of the forcing as described in Section 1.3. We found there that the complete (linear) solution, describing the growth of the storm surge from an ocean which is flat at time zero, can be described as the superposition of three waves:

$$\eta(x,t) = \eta_{\text{forced}}(x,t) + \eta^+_{\text{free}}(x,t) + \eta^-_{\text{free}}(x,t)$$

$$= A_{\text{forced}}f(x-ct) + A^+_{\text{free}}f(x-\sqrt{gh}\,t) + A^-_{\text{free}}f(x+\sqrt{gh}\,t) \quad (1.3.31)$$

The unsteady surface profile develops as the steady solution moves with the air pressure field at speed c while the two free waves move with speed $\sqrt{gh}$ in opposite directions.

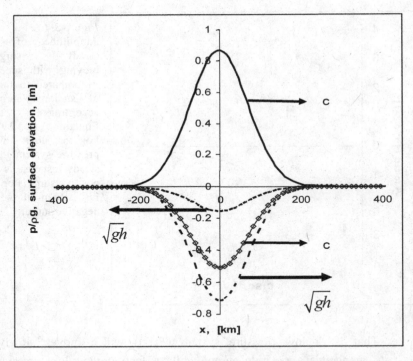

Figure 3.4.2: The initial flat surface is seen as the superposition of the steady, forced solution given by (1.3.29) and two free waves given by (1.3.43) which cancel it with respect to surface elevations and water velocities at $t=0$.

For $0<c<\sqrt{gh}$, the forward moving free wave increasingly dominates for $c\to\sqrt{gh}$ but both of (A_{free}^+, A_{free}^-) remain negative. For $c>\sqrt{gh}$, A_{forced} is negative, i e, the forced wave due to a fast moving low pressure system is a surface depression! Equation (1.3.33) now shows that the two free waves have opposite signs and A_{free}^+ dominates.

The development of the solution with time is shown in Figure 3.4.3 for the case of $c=0$, i e, a stationary pressure field switched on over a flat ocean at time 0.

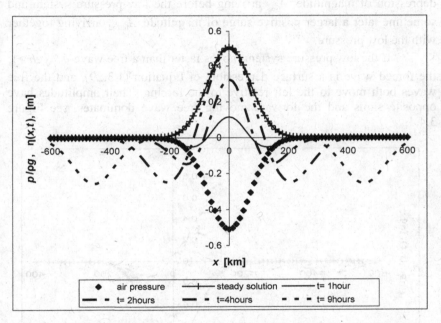

Figure 3.4.3: Sea surface shapes after 1 hour, 2 hours, 4 hours and 9 hours for a stationary low pressure system with peak $p/\rho g = -0.5$m, switched on over a calm sea, at $t=0$.

In Section 1.3.3, we found that, if the starting conditions are a flat surface and zero velocity everywhere, the amplitudes of the free waves are given by

$$(A_{free}^+, A_{free}^-) = \left(-\frac{A_{forced}}{2}[1+\frac{c}{\sqrt{gh}}], -\frac{A_{forced}}{2}[1-\frac{c}{\sqrt{gh}}]\right) \quad (1.3.43)$$

There is no friction so the free shallow-water-waves travel away with constant shape. In this case their speed is $\sqrt{gh}$ =14m/s. The forcing being stationary, the steady amplitude according to (1.3.19) is $-P_o/\rho g$. In this friction free case the free waves do not decay. Being shallow-water-waves and hence non-dispersive, they also do not disperse.

For moving low pressure systems in the range $0 < c / \sqrt{gh} < 1$, the forward moving free wave is largest: $A_{\text{free}}^+ > A_{\text{free}}^-$ as was shown in Figure 1.3.8. On a shoreline crossed by this system, an observer will see a surface depression of magnitude A_{free}^+ arriving before the low pressure system and sometime later a larger positive surge of magnitude A_{forced} arriving together with the low pressure.

If the low pressure system moves faster than a free wave $c / \sqrt{gh} > 1$, the forced wave is a surface depression, cf Equation (1.3.29), and the free waves both move to the left relative to the forcing. Their amplitudes have opposite signs and the forward moving free wave dominates, see Figure 3.4.4.

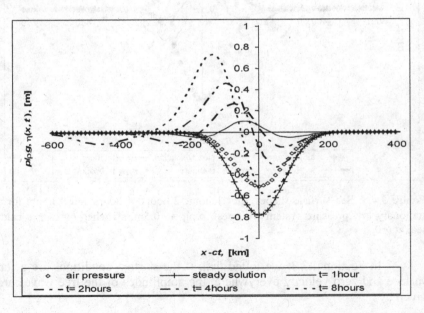

Figure 3.4.4: Same low pressure system as in the two previous figures but now moving faster than a free wave: c=18m/s while $\sqrt{gh}$ =14m/s. A_{forced}= −0.78m, $(A_{\text{free}}^+, A_{\text{free}}^-)$ = (0.89m, −0.11m).

The forward moving free wave now has a greater amplitude than the forced wave. Hence, if friction is negligible, so the free wave does not decay, the inundation sequence at landfall will be a negative surge of magnitude A_{forced} arriving together with the peak low pressure, followed by a larger positive surge of magnitude A_{free}^+. In other words, although the sign of A_{forced} changes from positive to negative at $c = \sqrt{gh}$, cf Figure 3.4.1, the qualitative experience at landfall does not change. It is always that of a negative surge followed by a larger positive surge, see Figure 3.4.5.

The scenarios in Figure 3.4.5 correspond to the forcing starting up abruptly. If the forcing is instead growing gradually in strength, η_{forced} is unchanged, but the free waves get stretched smeared out as explained in detail by Nielsen et al. (2008).

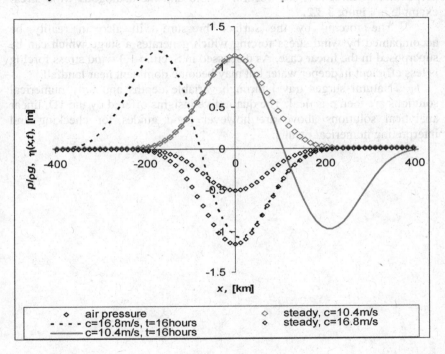

Figure 3.4.5: Landfall scenarios corresponding to the two marked cases in Figure 3.4.1 after a finite time so that the free waves are still around. Although the asymptotic steady solutions have different signs due to $c/\sqrt{gh}$ being on opposite sides of unity, the landfall scenarios are in both cases that of a negative surge followed by a larger positive surge.

At resonance, $c = \sqrt{gh}$, the steady η_{forced}-solution breaks down as $A_{\text{forced}}, A^+ \to \infty$ for $c \to \sqrt{gh}$. What happens then is that η_{forced} and η^+ merge into a linearly growing wave of the form $\eta_{\text{resonant}} \propto -t \dfrac{\partial}{\partial x} \mathrm{f}(x - ct)$, see Sections 1.3.3 and 2.3.6 for details. For example, a steady Gaussian-shaped forcing at resonance, drives a linearly growing N-wave, see the surf beat example in Section 2.3.6.

In reality, storm surges develop in two horizontal dimensions, where the $gh/(gh-c^2)$-amplification described by (1.3.29) is weaker. It is weaker because it only relates to the direction of travel of the forcing, while excess height, compared with the 1cm/hPa rule, radiates away in the perpendicular direction, see Section 1.3.3.6 and the analogous wind stress example in Figure 3.3.2.

The forcing by the surface pressure will also in reality be accompanied by wind stress forcing which generates a surge which can be superposed in the linear case. As discussed in Section 1.3 wind stress forcing is less efficient in deeper water but may become dominant near landfall.

Natural surges travel through variable depths and only numerical solutions are then practical. The qualitative insights offered by the 1D, linear analytical solutions above are however good guides for checking and interpreting numerical results.

4 Tides

4.1 THE NATURE OF OCEAN TIDES

Resulting from the varying strength of the gravitational pull from the Moon and the Sun the ocean levels around the world vary quasi periodically with periods of approximately 12.25 and/or 24.5 hours. The tides vary considerably from place to place with respect to both range and shape as indicated by Figure 4.1.1.

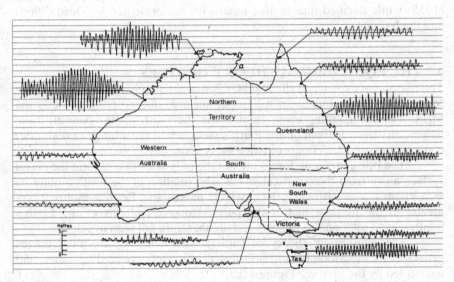

Figure 4.1.1: Recorded tides from January 1 to January 17, 1970, after Radock (1976).

The large tides that occur a day or two after full moon and new moon are called spring tides, while the minimal tides in between are called neap tides. Around the world the spring tidal ranges vary from zero to about 14m in funnel shaped estuaries like the Bay of Fundy, where the local topography enhances the tidal amplitude.

In some areas the tides are semidiurnal, i e, almost periodical with period 12.25 hours, in other areas they are diurnal with period 24.5 hours. These differences occur because the diurnal and semidiurnal tidal waves have different lengths and therefore generate different interference patterns when reflected between the continents.

The tidal currents are just as variable as the elevations that are illustrated in Figure 4.1.1. For example, the tidal currents are very weak along the southern part of the East Coast of Australia because the tidal waves in this area are much like standing waves of fairly uniform phase along the coast. Conversely, some of the strongest tidal currents (>2m/s) occur in the Torres Strait between Australia and New Guinea where the tide is more like a progressive wave.

A comprehensive, qualitative treatment of the tides is given by Defant (1958) while detailed quantitative treatments are presented by Dean (1966) and Pugh (2004).

4.2 THE EQUILIBRIUM THEORY OF TIDES

The reason we experience (lunar) tides is that the gravitational pull from the Moon creates two bulges in the ocean surface, one pointing towards the Moon and one pointing away from it. As the Earth rotates beneath these bulges, an observer fixed on the Earth's surface will see a bulge (a high tide) come past every 12.25 hours or twice every lunar day.

The tide generating forces are not the only ones that deform the ocean surface away from the spherical shape. The rotation of the Earth around its axis creates a centrifugal effect, which makes the Earth's radius 21km greater at the Equator that at the poles. However, effects of the Earth's rotation have rotational symmetry and are therefore not seen as tides.

The equilibrium theory of (lunar) tides describes the tidal bulges generated by the Moon on an Earth that is uniformly covered by water, and unaffected by the Sun, see Figure 4.2.1.

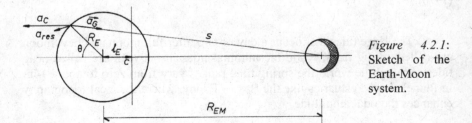

Figure 4.2.1: Sketch of the Earth-Moon system.

The mass of the Earth is $M_E = 6.0\times10^{24}$ kg, while that of the Moon is $M_M = 7.3\times10^{22}$ kg approximately 82 times less. The mean distance between the Earth and the Moon is $R_{EM} = 384,000$ km while the Earth's equatorial radius is $R_E = 6,378$ km. This means that the centre of mass C of the Earth-Moon system is inside the Earth, $l_E \approx 4,700$ km $\approx 0.74R_E$.

Since the orbital motion of the Earth and the Moon around C is stable, the centrifugal force and the gravitational pull from the Moon must be in balance at the centre of the Earth, i e,

$$\omega^2 l_E = G\frac{M_M}{R_{EM}^2} \tag{4.2.1}$$

where ω is the angular frequency and $G = 6.67$ Nm^2/kg^2 the gravitational constant.

As mentioned above, any effects of the Earth's rotation about its axis have rotational symmetry and are not experienced as tides. Hence, for the purpose of analysing tides, we may consider the forces acting on the surface of an Earth that is not rotating about its own axis while orbiting the common centre of mass. On such an Earth, all points experience the same centrifugal acceleration:

$$a_C \equiv \omega^2 l_E = G\frac{M_M}{R_{EM}^2} \tag{4.2.2}$$

However, the strength of the gravitational pull $\vec{a}_G$ from the Moon varies with the local distance s from the Moon as

$$|\vec{a}_G| = G\frac{M_M}{R_{EM}^2}\left(\frac{R_{EM}}{s}\right)^2 = G\frac{M_M}{R_{EM}^2}\frac{R_{EM}^2}{R_{EM}^2 + 2R_{EM}R_E\cos\theta + R_E^2}$$

$$\approx G\frac{M_M}{R_{EM}^2}\left(1 - 2\frac{R_E}{R_{EM}}\cos\theta\right) \tag{4.2.3}$$

In addition, $\vec{a}_G$ generally has a component perpendicular to the Earth-Moon centreline.

In a coordinate system with x-axis in the Earth-Moon direction we find

$$\vec{a}_{\text{res}} = \vec{a}_G - \vec{a}_C \approx G\frac{M_M}{R_{EM}^2}\begin{pmatrix} 1-2\,\dfrac{R_E\cos\theta}{R_{EM}}-1 \\ \dfrac{-R_E\sin\theta}{R_{EM}} \end{pmatrix} \qquad (4.2.4)$$

The component of this resultant in the direction of $\vec{g}$ is negligible compared with $\vec{g}$ itself. However, the component perpendicular to $\vec{g}$, i e, in the direction $\begin{pmatrix} -\sin\theta \\ \cos\theta \end{pmatrix}$ is effective. It will tilt the equilibrium ocean surface and give rise to the tide. Its magnitude is

$$a_{\text{eff}} = \frac{3}{2}G\frac{M_M}{R_{EM}^2}\frac{R_E}{R_{EM}}\sin 2\theta \qquad (4.2.5)$$

The effect of a_{eff} is to tilt the local equilibrium ocean surface the angle

$$\tan^{-1}\left(\frac{a_{\text{eff}}}{g}\right) \approx \frac{a_{\text{eff}}}{g} = \frac{3}{2}\frac{M_M}{M_E}\left(\frac{R_E}{R_{EM}}\right)^3\sin 2\theta \approx 8.4\times10^{-8}\sin 2\theta \qquad (4.2.6)$$

and thus create the bulges as shown in Figure 4.2.2.

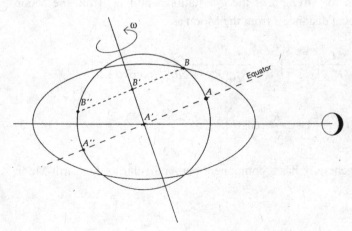

Figure 4.2.2: Variation in the Moon's gravitational attraction gives rise to two bulges centred on the Earth-Moon axis.

The peak of the bulges have a height of 36cm while the troughs are 18cm below the non-tidal ocean level giving a total "equilibrium" tidal range of 54cm. As the Earth rotates, observers fixed on the Earth will see tides related to these bulges.

When the Moon (or in general the axis of the bulges) is not in the plane of the Earth's Equator, the two daily high tides will have different sizes, except for observers on the Equator, see Figure 4.2.2. This gives rise to the diurnal tidal components.

The Sun generates a similar effect the magnitude of which is found by replacing M_M by the mass of the Sun ($M_s = 1.99 \times 10^{30}$kg) and R_{EM} by the mean Earth-Sun distance ($R_{ES} = 150,000,000$km) in the expressions above. That substitution shows that the tide generating forces of the Sun are of the order 45% of those of the Moon, in agreement with measurements.

The Sun and the Moon act together around full moon and new moon while their effects are opposed half way between full moon and new moon. This gives rise to the variation between spring and neap tides by a factor of the order $(1+0.45)/(1-0.45) \approx 2.6$, the monthly inequality, see Figure 4.1.1.

Further, small corrections occur because the orbits of the Earth and Moon are not exactly circular and because the direction of the Earth's axis is not fixed relative to the orbits of the Earth and the Moon.

4.3 MEASURED VERSUS PREDICTED TIDES

The action of the Sun and the Moon and the modifying effects due to the shapes of the oceans result in an, in principle, predictable periodic motion called the astronomical tide.

This astronomical tide is however not immediately measurable because the instantaneous, local sea levels also contain contributions from the changing weather patterns.

Figure 4.3.1 shows an example of a stationary weather effect which is being forced locally. A stationary weather system such as Cyclone Justin (Figure 4.3.1) can generate tidal anomalies through the winds and through the low atmospheric pressure as discussed in Sections 3.4 and 3.3.

Figure 1.3.3 shows Kelvin waves which are long period waves which travel around continents (clockwise in the northern hemisphere and anticlockwise in the southern hemisphere) after being generated by far-away weather systems.

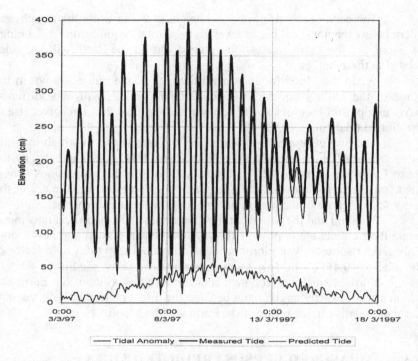

Figure 4.3.1: Recorded water levels, predicted astronomical tide and tidal anomaly at Cape Ferguson March 3 to 18, 1997. The large anomalies are caused by tropical Cyclone Justin.

4.4 PREDICTING THE ASTRONOMICAL TIDE

The pattern that results from the influence of the irregularly shaped continents on the tide is still too complicated to predict from basic principles. The astronomical tide predictions are therefore obtained by curve fitting to measured tide records. A procedure that, given long enough records, give accurate predictions.

While theoretical considerations still cannot predict the magnitude and phase of the various periodic components of the local tide, the periods can be predicted. The theoretical form of the local astronomical tide is therefore a series of the form

$$\eta_{tide} = \Sigma A_n \cos\left(\omega_n t - \phi_n\right) \tag{4.4.1}$$

where the angular frequencies ω_n or the periods T_n are theoretically determined, e g 12.42 hours for the semidiurnal lunar component M_2 and 12.00 hours for the semidiurnal solar component S_2. The local amplitudes A_n and φ_n phases are then determined by least squares fitting to the available local tide records. Modern tide predictions are often based on more than 40 constituents.

Since all the components have finite periods they will within a certain (long) time interval all be in phase and the highest astronomical tide given by

$$highest\ astronomical\ tide\ =\ \Sigma A_n \qquad (4.4.2)$$

would then occur in the absence of correlations and non-linear effects. In reality, negative correlations between for example M_1 and M_2 and friction effects mean that the value (4.4.2) would not be realised due to astronomical forcing alone.

4.5 TIDES IN RIVERS

The behaviour of the tide in rivers or long narrow estuaries depends on the effective tidal length L_{tide} of the river compared to the wave length of the tide: $T_{tide}\sqrt{gh}$ which is of the order 200km for typical depths of 2m. If the tidal part of the river is very long, the tidal wave will have the nature of a dampened progressive wave, which gradually vanishes before reaching the end. If the tidal river is short, the tidal wave becomes more like a standing wave with frictional damping.

4.5.1 The nature of river tides

In order to analyse the "long river" scenario, consider a shallow water sine wave, which decays exponentially due to friction:

$$\eta(x,t) = A_o e^{-k_r x}\cos(\omega t - k_i x) = A_o e^{-k_i x} Re\left\{ e^{i(\omega t - k_i x)}\right\} \qquad (4.5.1)$$

where A_o is the amplitude at $x = 0$, cf Figure 4.5.1. For this shallow water wave, the continuity equation (1.3.3) gives the horizontal water particle velocity

$$u(x,t) = \frac{A_o e^{-k_r x}}{h}\sqrt{gh}\cos\left[\omega t - k_i x + \tan^{-1}\left(\frac{k_r}{k_i}\right)\right] \qquad (4.5.2)$$

which, by comparison with the friction-free result (1.3.9), shows that the friction causes the local water velocities to lead the surface elevation by the phase angle $\varphi_f = \tan^{-1}\left(\dfrac{k_r}{k_i}\right)$ or by the time $\delta_{t,f} = \varphi_f / \omega$, see Figure 4.5.1.

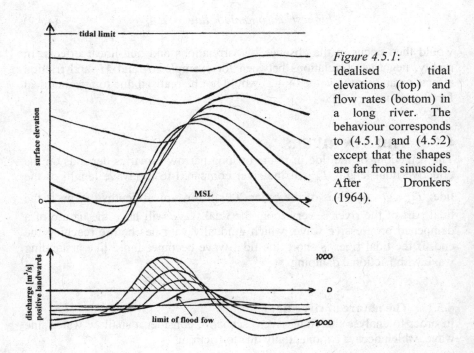

Figure 4.5.1: Idealised tidal elevations (top) and flow rates (bottom) in a long river. The behaviour corresponds to (4.5.1) and (4.5.2) except that the shapes are far from sinusoids. After Dronkers (1964).

The times of flow reversal are casually referred to as "slack water". For a friction-free progressive wave, they occur when $\eta = 0$. With friction, they occur $\delta_{t,f}$ earlier, e g, the reversal from inflow to outflow: "highwater slack," occurs after high tide but while the waterlevel is still above its local mean value. Correspondingly, "low water slack" occurs after low tide but before "mid tide." According to Equation (4.5.2) the delays of "high/low water slack" after high/low tide are the same, namely,

$T/4 - \delta_{t,f} = T/4 - T \tan^{-1}\left(\dfrac{k_r}{k_i}\right)/2\pi$. However, in most real cases the delay of low water slack is somewhat longer because of the finite amplitude effects, which steepen the front and flatten the back of the waves, c f Figure 1.5.2.

   If the river has significant freshwater flow so that the mean water level increases inland, the tide may* still be felt inland of the position where the mean water level equals high tide in the ocean. (*Provided the friction is not too strong and the river flow remains sub critical). For the case in Figure 4.5.1, the tidal limit is seen to be more than 1m above the high tide at the lowest stations.

   The limit of flood flow is the most inland point, where flow reversal occurs due to the tide. With zero freshwater flow it coincides with the tidal limit. For the case in Figure 4.5.1, it coincides with station 4.

   The attenuation factor k_r is, for shallow water waves, related to the friction coefficient f (defined by $\tau_{max} = \dfrac{1}{2} f \rho (u_{max})^2$ through

$$k_r \approx \frac{A_o}{3\pi h^2} f \qquad (4.5.3)$$

cf Nielsen (1992) p. 78.

   If the river is short compared with the tidal wave length and has little friction and a reflecting boundary at the upstream end, the seiche scenario of Figure 1.9.3 will occur. In the case of total reflection and no friction, velocities (positive inward) are 90 degrees ahead of the elevations. Hence, "slack water" occurs at the same times as high and low tides.

4.5.2   Numerical modelling of river tides

The tidal flow in rivers can be treated as finite amplitude, shallow water waves. Hence the governing equations are similar to the non-linear shallow water equations that were discussed in Section 1.5.2. However, in most cases it is warranted to include the extra complications due to friction, variable still water depth $d = d(x)$, and variable channel width $W = W(x)$.

   The momentum equation then becomes

$$\frac{\partial u}{\partial t} + u \frac{\partial u}{\partial x} = -g \frac{\partial \eta}{\partial x} - \frac{\tau}{\rho(d+\eta)} \qquad (4.5.4)$$

where the shear stress is usually written in the form $\tau = \frac{1}{2}\rho f |u| u$ leading to

$$\frac{\partial u}{\partial t} + u\frac{\partial u}{\partial x} = -g\frac{\partial \eta}{\partial x} - \frac{f}{2\rho(d+\eta)}|u|u \qquad (4.5.5)$$

The continuity equation becomes

$$W\frac{\partial \eta}{\partial t} = -\frac{\partial}{\partial x}\left[W(d+\eta)u\right] \qquad (4.5.6)$$

These equations are sometimes referred to as the Saint Venant equations.

They can be solved using the method of characteristics, but usually simplifications are worthwhile. For small Froude numbers, the fluid acceleration terms on the left of (4.5.5) may be neglected and a linear friction formula, $\tau = Ku$ or $\tau = KQ$ is usually introduced.

4.6   TIDES IN COASTAL LAGOONS

In tidal lagoons similar to the example in Figure 4.6.1 the tidal range may be greater or smaller than in the ocean depending on the overall geometry and on the roughness in the channel.

The dynamics can be modelled through the energy equation

$$\eta_o = \eta_B + \frac{L_C}{g}\frac{d\langle u\rangle}{dt} + \left[k_{\text{ent}} + k_{\text{ex}} + \frac{fL_C}{4h_C}\right]\frac{\langle u\rangle|\langle u\rangle|}{2g} \qquad (4.6.1)$$

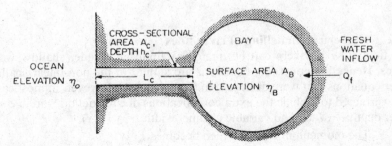

Figure 4.6.1: Definition sketch of a tidal lagoon. After Bruun (1968).

166

where the average landward velocity $<u>$ in the channel is related to the change in lagoon water level and the freshwater inflow Q_f by

$$\langle u \rangle = \frac{A_B}{A_C}\frac{d\eta_B}{dt_C} - \frac{Q_f}{A_C} \qquad (4.6.2)$$

so that the differential equation for the tide level in the bay becomes

$$\frac{L_C}{g}\frac{A_B}{A_C}\frac{d^2\eta_B}{dt^2} + \left[k_{ent}+k_{ex}+\frac{fL_C}{4h_C}\right]\left(\frac{A_B}{A_C}\right)^2\frac{1}{2g}\left|\frac{d\eta_B}{dt}-\frac{Q_f}{A_B}\right|\left(\frac{d\eta_B}{dt}-\frac{Q_f}{A_B}\right)+\eta_B=\eta_o$$

$$(4.6.3)$$

Due to the non-linearity of the friction term, Equation (4.6.3) does not have a simple analytical solution. The variation of A_C and h_C with the tide levels induce non-linear effects, which are clearly visible as a mean overheight (tidal pumping) of the lagoon above MSL even in the absence of freshwater flow and, a non-sinusoid shape of $\eta_B(t)$ even when the ocean tide is simple harmonic, cf Hinwood et al. (2005).

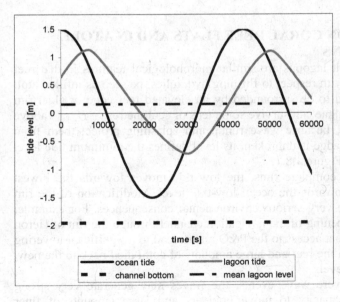

Figure 4.6.2: Tide in a coastal lagoon without freshwater flow. When the ocean tide range is large compared with the mean channel depth, the lagoon tide becomes quite asymmetrical and the mean lagoon tide becomes elevated above the MSL even if there is no freshwater flow.

In analogy with a linearly dampened oscillator such a lagoon system has the ability to resonate when friction losses are small. Resonance then occurs near the natural angular frequency of the undampened system, which is given by

$$\omega_0^2 = \frac{g}{L_C} \frac{A_C}{A_B} \qquad (4.6.4)$$

Several natural lagoons with tidal ranges greater than that in the adjacent ocean due to resonance have been recorded, see e g Bruun (1968) p. 97.

4.7   NET CIRCULATION DUE TO TIDES IN ESTUARIES

In a wide bay or estuary, the influence of the Coriolis effect and/or winds is often that the flood flow is concentrated along one side while the ebb flow is concentrated along the other. This means that, even though the nature of the tide is oscillatory with no net flow overall, there may well be a net circulation with the time mean velocity at most points being non-zero. This net circulation or different concentration of flood and ebb flows is usually also strongly indicated by the large scale bedforms.

4.8   TIDES ON CORAL REEF FLATS AND IN ATOLL LAGOONS

Reef flats and atoll lagoons are similar morphological features in that reef flats can, at least with respect to flushing hydraulics, be seen as infilled atoll lagoons. Both tend to be surrounded by an elevated reef edge or algal rim where corals and algae can survive some tens of centimetres ($= H_{\text{splash}}$) above the low-tide level, because wave runup and splashing protect them from drying out. If this edge is unbroken, its level defines the minimum tide level in the lagoon, see Figure 4.8.1.

With less complete rims, the low tide moves towards the lowest outflow level or towards the ocean low-tide level. Modification of the rim can therefore have very serious environmental consequences. For example, widening and deepening of the navigation channel through the rim of Heron Reef to improve boat access to the P&O resort, lead to a significant lowering of the low tides on the reef and large amounts of coral died back to the new, reduced low-tide level.

During severe wave events, the waves may generate very strong currents with a capacity to move boulders and large amounts of finer

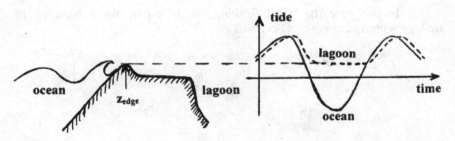

Figure 4.8.1: If the tidal range is large compared with H_{splash}, the high tide will be well above the reef edge and the flood flow will enter from all directions, perhaps with some directional bias due to wind stress and/or wave thrust. The low-tide level will be set by the reef edge.

sediments, Gourlay and Colleter (2005). However, the high-tide level in the lagoon will usually not be significantly above the ocean high-tide level.

Where the tidal range is small, i e, $< H_{splash}$, the lagoon tides will tend to be well above the ocean tides if the rim is complete. This is the case for the Manihiki atoll, where the lagoon tide is above the ocean tide at all times, see Figures 4.8.1 and 4.8.2.

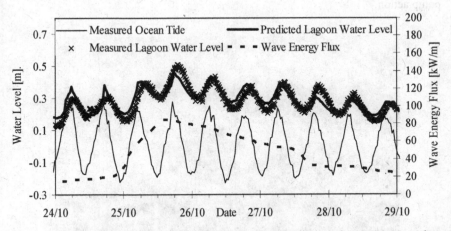

Figure 4.8.2: Ocean and lagoon tides at the Manihiki atoll, October 2004. Note that the lagoon water levels rise during periods with larger wave energy flux, indicating the wave driven nature of the flushing. After Callaghan et al. (2006).

In this case the "tidal flushing" is driven by wave pumping in analogy with rip currents, cf Section 2.6.

Figure 4.8.3a: On the side facing the wave action, water enters the lagoon by wave pump action.

Figure 4.8.3b: Outflow, on the side with no waves: The water escapes to the ocean on the right.

5 Estuaries and estuarine hydraulics

5.1 INTRODUCTION

Broadly defined, estuaries are the areas of interaction between fresh water from precipitation on land and salt water from the open oceans. As a result estuarine waters cover the salinity range from zero to 35 g/l and density range from about 999 to around 1030 kg/m^3 (depending on temperature).

The flows, which will tend to mix the salty and fresh water, may be due to steady river discharge, tidal flows, wind waves or circulation driven by wind or density differences. The mixing will be impeded by buoyancy effects which can be quite strong, e g, a balloon filled with fresh water and submerged in sea water will deliver 25 times the buoyancy force of a same size helium balloon in air, cf Figure 5.3.1.

5.2 CLASSIFICATION OF ESTUARIES

Estuaries can be classified according to either geomorphology or salinity structure. For more details see e g, Dyer (1997), Open University (1999) or Savanije (2005).

5.2.1 Estuary geomorphology

At one end of the morphological spectrum are the **fjords** of Norway, New Zealand and Chile. These are deep valleys scoured by the ice, possibly all the way down to the sea level during the last glaciation, about 120m below the present sea level. The cross sections are U-shaped and they contain little sediment.

The next geomorphological family is **drowned river valleys**. As the name indicates these are river valleys formed while the sea level was considerably lower than at present. They have since been drowned due to rising sea levels and the flow has become more or less dominated by tides. As the sedimentation progresses drowned river valleys may become restricted by a bar at the ocean, they are then classified as **bar built estuaries**. In some cases these can become totally closed by the bar between major freshwater flood events.

171

5.2.2   Classification by salinity structure

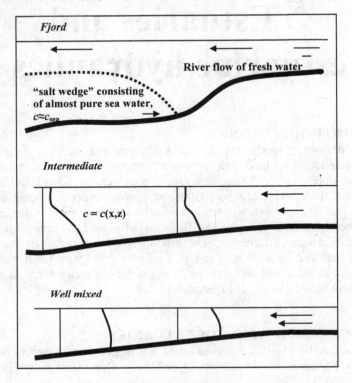

Figure 5.2.1: Estuaries classified by salinity structure. Top: Fjords, there is little vertical mixing, i e, almost pure fresh water is flowing out on top of ocean water. These are also called salt wedge estuaries. Middle: Partly mixed estuary. In each vertical there is a gradual change in salinity rather than a sharp interface. At a given depth, the salinity decreases gradually in the landward direction. Bottom: Well mixed estuary. Due to strong vertical mixing, the salinity varies little over the depth.

5.3   DENSITY DRIVEN FLOWS AND STRATIFICATION

5.3.1   Density variation

The density of ocean and estuarine waters is a function of both temperature and salinity (salt concentration) as indicated by Table 5.3.1.

Table 5.3.1: Density [kg/m³] of water as function of temperature and salinity at near atmospheric pressure.	Temperature [°C]	Fresh water	Sea water (35ppt)
	0°	999.8	1028.1
	30°	995.8	1021.8

The range of densities occurring within the "normal" variability of coastal and estuarine conditions is thus about 996 to 1028 kg/m³ corresponding to a maximal relative density difference of $\Delta_\rho/\rho = 32/996 = 0.032$. This can give rise to very significant buoyancy forces and buoyancy driven flows. A few "negative" estuaries with more evaporation than freshwater inflow become hypersaline and therefore have greater density than the bordering ocean.

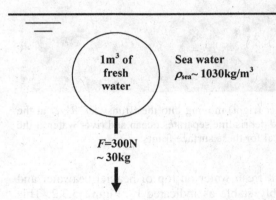

1m³ of fresh water

Sea water $\rho_{sea} \sim 1030 kg/m^3$

F=300N ~ 30kg

Figure 5.3.1: A force of 300N is required in order to submerge 1m³ of fresh water in the sea. Helium balloons in air provide only around 1/30 of this lift per unit volume. The strength of this buoyancy makes diffuser design very important for ocean sewage outfalls.

The heating of the oceans in equatorial areas and cooling in the polar areas gives via the ensuing density differences rise to large ocean circulation systems including the Gulf Stream and the East Australian Current who are both carrying warm equatorial surface water towards the poles.

Other ocean currents are driven predominantly by salinity differences. Thus, the Mediterranean Sea, which loses more water by evaporation than it receives from rain and rivers, is saltier than the Atlantic. As a result, there is a strong flow of Atlantic surface water through the Straits of Gibraltar partly compensated by an opposite flow of Mediterranean bottom water (cf Figure 5.3.3 for the two-way forcing).

On a smaller scale, salty ocean water is brought face to face with more or less fresh river water when the tide starts to flood as in Figure 5.3.2.

Figure 5.3.2: Clear salty sea water (right) moving into the Brunswick River at the start of the flood tide. A foam and debris line separates ocean and river water at the surface. The wedge shape is typical for these surface fronts.

Stratification with light fresh water on top of heavier seawater and associated fronts are remarkably stable as indicated by Figure 5.3.2. This front has remained sharp after moving through the breaker zone over the entrance bar to the left of the photo. In other words, the mixing of stratified fluid requires considerable work. For example, the mixing of $h/2$ of fresh water on top of $h/2$ of sea water, Figure 5.3.3, into h of uniform density $(\rho_{fresh}+\rho_{sea})/2$ requires $(\rho_{sea}-\rho_{fresh})gh^2/8$ [J/m^2].

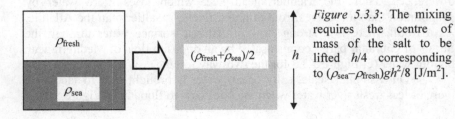

Figure 5.3.3: The mixing requires the centre of mass of the salt to be lifted $h/4$ corresponding to $(\rho_{sea}-\rho_{fresh})gh^2/8$ [J/m^2].

On the global scale of the Earth's oceans, this kind of mixing of warm, hence light, ocean surface water with cooler, denser underlying water consumes around 3TW of power, cf St Laurent and Simmons (2006).

The equivalent freshwater outflow corresponding to the flow of brackish water with salinity $c(z)$ is

$$Q_{\text{fresh}} = \int \frac{c_{\text{sea}} - c}{c_{\text{sea}}} dq = \int\limits_{\text{bottom}}^{\text{surface}} \frac{c_{\text{sea}} - c(z)}{c_{\text{sea}}} u(z) B(z) dz \qquad (5.3.1)$$

where $B(z)$ is the channel width.

5.3.2   Flow forcing from density differences

Imagine a situation where the light (density ρ) fresh water on the left and the sea water ($\rho + \Delta_\rho$) on the right of Figure 5.3.2 were separated by a vertical membrane as in Figure 5.3.4.

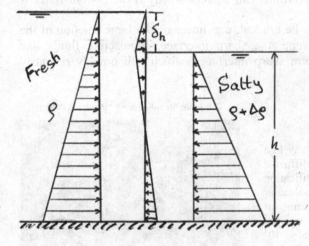

Figure 5.3.4:
The fresh water on the left is in overall force balance with denser salt water on the right. However, local pressures are not balanced. The resultants are shown in the middle.

An overall pressure-force balance between the two sides would then require an overheight δ_h, of the lighter water, given by

$$\frac{1}{2}\rho g (h + \delta_h)^2 = \frac{1}{2}(\rho + \Delta_\rho) g h^2 \qquad (5.3.2)$$

175

and hence

$$\delta_h \approx \frac{1}{2}\frac{\Delta_\rho}{\rho}h \qquad (5.3.3)$$

where h is the depth of salt water. It is clear that, if the vertical membrane is removed, fresh water will move to the right on top while salt water will move towards the left at the bottom. This motion, moderated by friction, would in the absence of mixing and other forcing, continue until the fresh water comes to rest on top of the salt water with a horizontal interface.

5.3.3   Stratification and restratification

Layered or stratified flows are common in the oceans and estuaries as well as in the atmosphere. Stratifications may be stable or unstable. They are stable when lighter fluid flows on top of heavier fluid. Unstable stratification and eventual overturns are sometimes observed in lakes and canals during winter, when cooling at the surface can make the surface water denser than the bottom water. Such overturns can be very smelly if the bottom water is rich in H_2S.

Stratification may be gradual, e g, linear over a large fraction of the depth, or it may be abrupt at a sharp interface. Immiscible fluids like kerosene above water form sharp interfaces, which will quickly restratify after stirring.

Figure 5.3.5:
Up-gradient transfer of density in a thermo-haline, double diffusive system. This restratification mechanism can, at the molecular level, be understood in terms of differential diffusion i e, the molecular diffusivity of heat is 10^4 times greater than that of salt. It is however not obvious why such differences should exist between the corresponding turbulent diffusivities. Source: SCOR working group on oceanographical double diffusion.

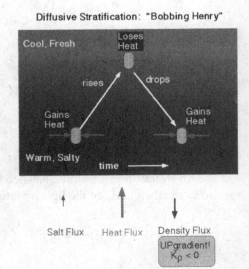

176

Some restratification may also occur between layers of water with different density when the molecular diffusion is insufficient for evening out the density of entrained blobs before they can be returned by gravity.

In the somewhat surprising, but nevertheless, naturally occurring scenario of cool fresh water above warm salty water (Figure 5.3.5) the restratification effect is particularly strong. In this scenario, the diffusion of density is in fact "up-gradient" corresponding to a negative diffusivity for density: $K_\rho < 0$. The density gradient is thus unstable (grows unboundedly) and the supply of heat from below drives a sharpening of the interface. This instability of the density gradients results in the so-called thermo-haline staircase, which is commonly observed in the oceans. That is the density variation with depth is often not smooth, but displays a number of sharp steps.

5.3.4 Stratified fluids in motion

In Figure 5.3.2, the fresh water overheight is less than the equilibrium value (5.3.3) and, consequently, the front moves up-river (to the left). A general vertical section through such a *salt wedge* system is shown in Figure 5.3.6. The surface and the salinity are both continuous in natural systems, nevertheless, the nature of the forcing of density driven circulation is still as indicated by Figure 5.3.4. That is, horizontal density gradients, or the slope of a density interface, generate horizontal pressure gradients which are analogous to the residual pressures in the centre of Figure 5.3.4.

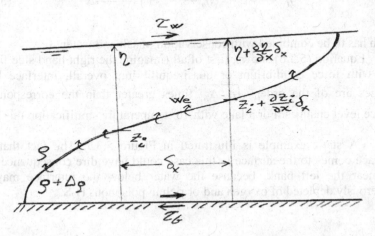

Figure 5.3.6: A general salt wedge system with interface height $z_i(x)$ and entrainment velocity w_e. The interface meets the bed at the *null point* where velocity and bed shear stress both vanish for a stationary wedge.

The steepness of the interface is related to the overall speed of the system. A fast moving system, like the one in Figure 5.3.2, will in general have a fairly steep interface while stationary systems, e g, quiescent lakes with temperature stratification, have horizontal interfaces. A famous, almost stationary system is that at Gibraltar, where tidal changes are moderate compared with the mean flow pattern. For a good discussion of the basic dynamics see Keulegan (1966) or Wilson (1988).

An idea about what determines the shape of the interface can be obtained by assuming overall force equilibrium for a vertical slice of width δ_x, cf Figure 5.3.6.

$$\frac{1}{2}\rho g \eta^2 + \frac{1}{2}\Delta_\rho g z_i^2 + \tau_w \delta_x = \frac{1}{2}\rho g \left(\eta + \frac{\partial \eta}{\partial x}\delta_x\right)^2 + \frac{1}{2}\Delta_\rho g \left(z_i + \frac{\partial z_i}{\partial x}\delta_x\right)^2 + \tau_b \delta_x$$

(5.3.4)

$$\eta \frac{\partial \eta}{\partial x} + \frac{\Delta_\rho}{\rho} z_i \frac{\partial z_i}{\partial x} \approx \frac{\tau_w - \tau_b}{\rho g}$$

(5.3.5)

which can be integrated, for example with boundary conditions $\eta(0) = \eta_o$ and $z_i(0) = z_o$ to give

$$\eta^2 + \frac{\Delta_\rho}{\rho} z_i^2 = \eta_o^2 + \frac{\Delta_\rho}{\rho} z_o^2 + \int_o^x \frac{\tau_w - \tau_b}{\rho g} dx$$

(5.3.6)

which has to be combined with conservation of salt and water.

Equation (5.3.6) shows, first of all (imagine the right hand side fixed) that, with force equilibrium or quasi-equilibrium overall, interface level changes are of the order $\sqrt{\rho/\Delta_\rho}$ times greater than the corresponding surface level changes. For a lake with $10°$ temperature stratification this ratio is 27.

A static example is illustrated in Figure 5.3.7. The fact that the interface comes to the surface in this case could have dire consequences for fish near the left bank, because the water below the interface may be dangerously depleted of oxygen and/or rich in poisonous H_2S.

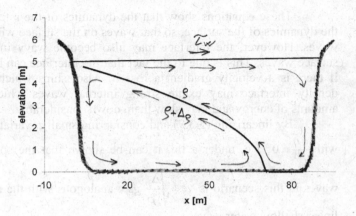

Figure 5.3.7:
Wind stress towards the right makes the lake surface slope upwards to the right and the interface slope downwards to the right.

For a system with no mixing across the interface, the dynamics can be investigated by applying the continuity equation to both layers together with Newton II which, noting that the pressure at a level z below the interface ($z < z_i$,) is $p(z) = \rho g(\eta - z) + \Delta_\rho g(z_i - z)$, gives

$$\frac{du}{dt} = -g\frac{\partial \eta}{\partial x} + \frac{1}{\rho}\frac{\partial \tau}{\partial z} \qquad \text{for } z > z_i \qquad (5.3.7)$$

$$\frac{du}{dt} = -g\left(\frac{\partial \eta}{\partial x} + \frac{\Delta_\rho}{\rho}\frac{\partial z_i}{\partial x}\right) + \frac{1}{\rho}\frac{\partial \tau}{\partial z} \quad \text{for } z < z_i \qquad (5.3.8)$$

where the approximation $\dfrac{\rho + \Delta_\rho}{\rho} \approx 1$ has been applied. The approximate dynamics of the interface may be determined by integrating (5.3.8) over the bottom layer, assuming uniform velocity, and combining with the continuity equation for the lower layer, i e,

$$z_i\frac{du}{dt} = -gz_i\left(\frac{\partial \eta}{\partial x} + \frac{\Delta_\rho}{\rho}\frac{\partial z_i}{\partial x}\right) + \frac{\tau_i - \tau_b}{\rho}$$

$$\frac{\partial z_i}{\partial t} = -\frac{\partial}{\partial x}(u z_i) \qquad (5.3.9)$$

These equations show that the dynamics of the interface are tied to the dynamics of the surface, so that waves on the surface will drive interface waves. However, the interface may also become wavy in the absence of surface waves. This is due to the fact that the interface can become unstable if there is a velocity gradient across it. Also, ships which travel above a density interface may excite strong internal waves which extract large amounts of energy and may slow them down considerably.

By linearising (5.3.9) and considering small z_i-variations in a system with $\dfrac{\partial \eta}{\partial x} \equiv 0$, i e, under a lid, it can be shown that the speed of interface waves in this scenario is $c_i = \sqrt{\dfrac{\Delta_\rho}{\rho} gz_i}$, analogous with the result (1.3.5) for linear shallow water waves.

In analogy with free surface flows, the dynamics of flows with typical velocity V and involving an interface with density difference Δ_ρ can be classified in terms of the *densimetric Froude number*

$$F_{\Delta_\rho} = \frac{V}{c_i} = \frac{V}{\sqrt{\dfrac{\Delta_\rho}{\rho} gz_i}} \tag{5.3.10}$$

5.3.5 Shear instability at an interface
Internal waves at a density interface, like surface waves, and like waves between layers of different density in the atmosphere are usually initiated by shear instability or Kelvin Helmholtz instability, the essence of which is that flow along a curved wave crest generates a centrifugal force and hence suction, see Figure 5.3.8. If this suction is strong enough to lift the wave

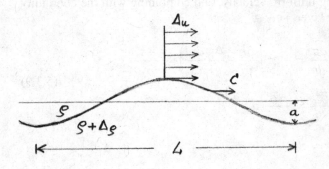

Figure 5.3.8: Depending on the relative magnitude of the density difference Δ_ρ and the relative velocity Δ_u an initial deformation of the interface may grow or decay.

180

crest up even higher the waves will grow, and the interface is unstable, otherwise they will decay and the interface is stable.

To find a simple stability criterion, we can assume that the bottom layer is at rest while the upper layer moves with uniform velocity Δ_u. Following the curvature of the crest of the simple sine wave $\eta(x,t) = a\cos(\omega t - kx)$, with relative velocity $\Delta_u - c$ requires the centripetal acceleration $-k^2 a(\Delta_u - c)^2$ which, acting through a layer of thickness k^{-1} (compare the pressure distribution under a deep water wave), generates a suction pressure of magnitude $\rho ka(\Delta_u - c)^2$. The suction required to hold the lower fluid elevated to the crest height a is $\Delta_\rho ga$. Equating these two pressures gives the criterion for instability

$$ k > \frac{\dfrac{\Delta_\rho}{\rho}g}{(\Delta_u - c)^2} \tag{5.3.11} $$

i e, the stronger the stratification, the shorter the waves that can be generated. For a more comprehensive analysis, which also gives the rate of growth and the wave speed, see Liggett (1994), p. 216. For the analogous initiation of surface wind waves, the surface tension enters the equations, cf Kinsman (1965) or Phillips (1966).

When the instability is due to the Kelvin Helmholtz mechanism, the main governing parameter is the overall Richardson number Ri_o. Instabilities occur when

$$ Ri_o = \frac{\dfrac{\Delta_\rho}{\rho}g}{\Delta_u^2 / L_u} < 0.25 \tag{5.3.12} $$

where L_u is the vertical scale of the velocity variation.

5.3.6 Mixing or entrainment across an interface

For immiscible fluids like water and kerosene there is no transport of mass across the interface and continuity can be applied to each layer separately. In natural systems however, there is usually a certain amount of mixing at the interface. If the top layer is much more turbulent than the bottom layer, this mixing is essentially a one-way process with entrainment of the denser bottom water into the upper layer, leading to a gradual increase of the density in the upper layer, while the density in the quiescent lower layer remains virtually unchanged. For an introduction, see Turner (1973), Chapter 9.

The rate of mixing across an interface, is usually quantified by the entrainment velocity w_e. With a few simplifying assumptions, w_e can be determined by balancing the potential energy per unit time and area required for this mixing rate:

$$\dot{E}_p = \Delta_\rho g w_e \tag{5.3.13}$$

and the local rate of production of turbulent kinetic energy $\dot{E}_t$ at the interface. The latter is given by

$$\dot{E}_t = \tau \frac{\partial u}{\partial z} = \rho v_t \left(\frac{\partial u}{\partial z} \right)^2 \tag{5.3.14}$$

See e g, Landau and Lifschitz (1987) p. 50 or Batchelor (1967) p. 152.

Assuming $\dot{E}_p \approx \dot{E}_t$ then leads to an expression for the entrainment velocity

$$w_e \approx \frac{\rho v_t \left(\frac{\partial u}{\partial z} \right)^2}{\Delta_\rho g} \tag{5.3.15}$$

However, the eddy viscosity v_t is strongly influenced by the stratification, so this is not a neat closed solution. The problem is thus a very complex one, which cannot be resolved to great accuracy by purely theoretical means. In addition, the density changes are often gradual rather than stepwise. The experimental data is so far also quite inconclusive, see e g, Fernando (1991) for a review.

The mass flux density $w_e \Delta_\rho$ may alternatively be evaluated as $\overline{w' \rho'}$ or in terms of a diffusivity K_ρ as $-K_\rho \frac{\partial \rho}{\partial z}$. Barry et al. (2001) studied the diffusivity of mass K_ρ in a stirred tank where the mixing is driven by an oscillating grid rather than by interface shear. They identified a weak and an energetic regime in which K_ρ can be determined from

$$
K_\rho = \begin{cases} 0.9 v^{2/3} \kappa^{1/3} \dfrac{\dot{E}_t}{v \dfrac{g}{\rho} \dfrac{d\rho}{dz}} & \text{for } 10 < \dfrac{\dot{E}_t}{v \dfrac{g}{\rho} \dfrac{d\rho}{dz}} < 300 \\[4em] 24 v^{2/3} \kappa^{1/3} \left(\dfrac{\dot{E}_t}{v \dfrac{g}{\rho} \dfrac{d\rho}{dz}} \right)^{1/3} & \text{for } 300 < \dfrac{\dot{E}_t}{v \dfrac{g}{\rho} \dfrac{d\rho}{dz}} < 10^5 \end{cases}
\tag{5.3.16}
$$

where κ is the thermal conductivity of the fluid.

A related quantity to the mass flux density is the buoyancy flux (density) $-g \overline{w'\rho'} / \rho$ which has dimension of power per unit mass. It may be seen as the rate at which gravitational energy is released per unit mass, by downward transfer of heavier fluid.

The overall nature of the mixing process is indicated by a Richardson number which gives the ratio between available turbulent mixing energy and the required potential energy: One type of Richardson number is the gradient Richardson number

$$
Ri = \frac{g \dfrac{\partial \rho}{\partial z}}{\rho \left(\dfrac{\partial u}{\partial z} \right)^2} \propto \frac{\dot{E}_p}{\dot{E}_t}
\tag{5.3.17}
$$

Here, the strength of the stratification is represented by $\dfrac{g}{\rho} \dfrac{\partial \rho}{\partial z}$. The corresponding frequency

$$
N = \sqrt{\frac{g}{\rho} \frac{\partial \rho}{\partial z}}
\tag{5.3.18}
$$

is called the *stratification frequency, the buoyancy frequency* or the *Brunt-Väisälä frequency*.

For more details on mixing through interfaces, see e g, Wüest and Lorke (2003), Peltier and Caulfield (2003) or Ivey and Imberger (1991).

5.3.7 Turbidity currents and saline underflows

Turbidity currents and saline underflows are flows of a relatively dense bottom layer under an essentially quiescent upper water body, Figure 5.3.9. In turbidity currents, the increased density of the bottom layer may be due to suspended sediment. In saline underflows it is due to dissolved minerals (salt). Self sustaining turbidity currents were first described by Bagnold (1962).

Turbidity currents may become very violent where sediment has built up over a long time to a near limiting slope and then gets disturbed, for example by an earthquake. Rupture of marine cables have been attributed to such currents, cf Krause et al. (1970). The dynamics of these violent systems are described in detail by Parker (1982) and by Parker et al. (1986). For an example of the less violent but nevertheless very important saline underflows, see e g, Dallimore et al. (2001).

For a simplified scenario with uniform density $\rho + \Delta_\rho$ and velocity U in the bottom layer, the turbidity current can be described by the following set of equations: The continuity equation

$$\frac{\partial z_i}{\partial t} = -\frac{\partial}{\partial x}(z_i U) + w_e \qquad (5.3.19)$$

the momentum equation

$$(\rho + \Delta_\rho)\left(\frac{\partial}{\partial t}(Uz_i) + \frac{\partial}{\partial x}(U^2 z_i) + w_e U\right) = -\frac{\partial}{\partial x}(\frac{1}{2}\Delta_\rho gz_i^2) - \Delta_\rho gz_i S_b - \tau_b \qquad (5.3.20)$$

and the density equation

$$\frac{\partial}{\partial t}(\Delta_\rho z_i) = -\frac{\partial}{\partial x}(\Delta_\rho z_i U) + \rho_s(p - d) - w_e \Delta_\rho \qquad (5.3.21)$$

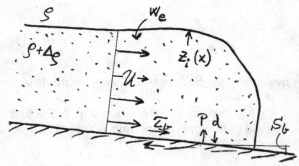

Figure 5.3.9:
Turbidity current on a sloping seabed. Sediment of density ρ_s may be picked up at the rate p and/or deposited at the rate d [LT^{-1}].

5.4 DISPERSION PROCESSES

Estuaries are the scene of a great variety of mixing and dispersion processes. Salty seawater is mixed with fresh river water resulting in a distribution of salt concentrations or salinities. Wastewater released into the estuary will gradually be dispersed and sediment suspended from the banks and bottom will be mixed through the water column under energetic conditions and settle out again at slack water.

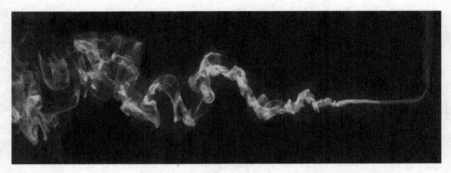

Figure 5.4.1: Flourescein dye dispersing from a 1.5mm tube issuing into a 10cm deep stream flowing at 1cm/s from right to left. There is mixing at three scales, the molecular scale, the scale of the pipe diameter and the scale of the water depth.

In Figure 5.4.1 the mixing plume is bounded by the flume bed and by the surface. In a stratified waterbody, the stratification may likewise inhibit dispersion.

Many books have been written on environmental dispersion processes. For example, Lewis (1997) covering estuaries and coastal waters, Rutherford (1994) covering rivers in general, Chanson (1997) covering air bubble entrainment and Wood et al. (1993) covering ocean disposal of wastewater. A classical text with a thorough mathematical coverage of environmental mixing and dispersion problems is Fisher et al. (1979), and a recent review with comprehensive bibliography is Wüest and Lorke (2003).

The behaviours of the solutes are diverse. The simplest are those that are not chemically reactive and have negligible settling (or rise) velocity. This can usually be considered to be the case for sea salt, *HCl* and for clay sediments. For non-reactive solutes, the total mass does not change with time, and for solutes with zero settling velocity the centre of mass is stationary when there are no net currents. Only non-reactive solutes are covered in the following.

5.4.1 Concentration distributions and their statistics

Concentrations of solutes can be given in different units, e g, moles per litre, grams per litre (g/l) or parts per million (ppm) by volume or by mass. In the following we use the notation $C(x,y,z,t) = C(\vec{r},t)$ for concentration and when nothing else is mentioned it is, the dimensionless volume/volume concentration.

The total (concentrated) volume of solute is calculated as

$$V_o(t) = \int_{\text{domain}} C(\vec{r},t)\,dV \qquad (5.4.1)$$

and if the domain has a finite volume V_{domain} the average concentration is defined as: $<C(t)> = V_o(t)/V_{\text{domain}}$. For non-reactive solutes in domains with no sources or sinks neither V_0 nor $<C>$ are functions of time.

The solute distribution has a centre of mass with position vector:

$$<\vec{r}(t)> = \frac{1}{V_o} \int_{\text{domain}} \vec{r} C(\vec{r},t)\,dV \qquad (5.4.2)$$

and the spread or standard deviation σ given by

$$\sigma^2(t) = \frac{1}{V_o} \int_{\text{domain}} (\vec{r}-<\vec{r}>)^2 C(\vec{r},t)\,dV \qquad (5.4.3)$$

Both of these quantities have directional components, i e, $<\vec{r}(t)> = (<x>,<y>,<z>)$ and $\sigma^2 = \sigma_x^2 + \sigma_y^2 + \sigma_z^2$.

5.4.2 Flux and flux density

The flux Q through a given surface of a substance or property ($\bullet$) which is distributed with density (or concentration) $c_\bullet(x,y,z)$ and which is being advected by the velocity field $\vec{u}(x,y,z)$ is quantified as

$$Q = \int_{\text{surface}} c_\bullet(\vec{u}\cdot\vec{n})\,dA \qquad (5.4.4)$$

where $\vec{n}$ is a unit vector normal to the surface. For the case of volume flux $c_\bullet=1$, for mass flux $c_\bullet$ is the usual mass density ρ and for sediment or solute

$c_{\bullet}$ is the concentration. We can also talk about flux of vector quantities, e g momentum $\rho \bar{u}$, for which we get

$$\overline{\text{momentum flux}} = \int_{\text{surface}} \rho \bar{u} (\bar{u} \cdot \bar{n}) dA \qquad (5.4.5)$$

Correspondingly, the flux of x-momentum, ρu, through a horizontal area is $\int_{\text{surface}} \rho u w dA$. In each case, the integrand is the flux density.

For turbulent flows we usually write the velocity as the sum of the time mean velocity $\bar{u} = (\bar{u}, \bar{v}, \bar{w})$ and the fluctuations $\bar{u}' = (u', v', w')$ and similarly for mass density, temperature and the concentrations of solutes or sediments. In this scenario the fluxes and flux densities will have contributions due to the time-mean and the fluctuations, which we often consider separately. That is for the flux density of solute or sediment through a horizontal plane we write

$$\overline{[\bar{c} + c'][\bar{w} + w']} = \overline{\bar{c} \, \bar{w}} + \overline{\bar{c} \, w'} + \overline{c' \, \bar{w}} + \overline{c' w'} = \bar{c} \, \bar{w} + 0 + 0 + \overline{c' w'} \qquad (5.4.6)$$

and similarly for the flux density of x-momentum upwards through a horizontal plane:

$$\overline{[\rho \bar{u} + \rho u'][\bar{w} + w']} = \rho \bar{u} \, \bar{w} + \rho \overline{u' w'} \qquad (5.4.7)$$

where the last term is minus the familiar turbulent Reynolds stress. In other words, in the usual notation, shear stress in general and Reynolds stress in particular is downward flux density of x-momentum.

5.4.3 Advective dispersion

The solute distribution and its statistics will change with time in a fashion that depends on the nature of the flow. A fixed amount of solute in a steady uniform flow will be advected downstream without changes to the shape of the distribution, but in general, the non-uniformities of the flow field will deform the distribution as illustrated by the following example.

Example 5.4.1:
Consider a circular pipe of radius R with laminar flow in the x-direction. Because of the axial symmetry we have $C = C(x, r, t)$, where r is distance from the centreline. At time zero dye is injected uniformly across the pipe cross section at $x = 0$, i e,

187

$$C(x,r,0) = \frac{V_o}{\pi R^2 \delta_x} \tag{5.4.8}$$

where δ_x is the infinitesimal thickness of the dye injection plane.
The velocity distribution is given by

$$u = u(r) = U\left[1 - \left(\frac{r}{R}\right)^2\right] \tag{5.4.9}$$

which corresponds to the mean velocity $\langle u \rangle = U/2$. As time passes, the initially plane dye surface is then deformed into a rotational paraboloid with apex at $x=Ut$.

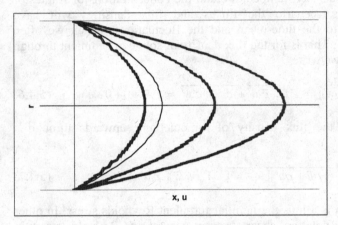

Figure 5.4.2:
Velocity distribution (thin line) and dye distributions at times 0.8s, 1.6s and 2.4s in a non-diffusive Poiseulle flow in a pipe.

Let us first determine the behaviour of the centre of mass of the dye distribution as time progresses. Because of the circular symmetry, the centre of mass remains on the centreline ($r=0$), and the problem of determining its coordinates is simply to determine $\langle x \rangle$ averaged over all dye particles. If there is no diffusion, the volume of dye, which is initially contained in a belt of width δ_r at radius r, will remain within the corresponding cylinder shell, but at time t it will be at $x= u(r)t$. In analogy with (5.4.2) we find

$$\langle x(t) \rangle = \frac{1}{V_o}\int xC\,dV = \frac{1}{V_o}\int_o^R U\left[1 - \left(\frac{r}{R}\right)^2\right]t\,\frac{V_o}{\pi R^2 \delta_x}2\pi r\delta_x\,dr$$

$$= \frac{1}{2}Ut = \langle u \rangle t \tag{5.4.10}$$

188

i e, the centre of mass moves along at the mean flow velocity $<u> = U/2$. The spread in the x-direction is found in analogy with (5.4.3) to be:

$$\sigma_x^2(t) = \frac{1}{V_o} \int (x - <x>)^2 C\, dV$$

$$= \frac{1}{V_o} \int_0^R \left\{ U \left[1 - \left(\frac{r}{R} \right)^2 \right] t - \frac{1}{2} Ut \right\}^2 \frac{V_o}{\pi R^2 \delta_x} 2\pi r \delta_x\, dr$$

$$= \frac{1}{12} (Ut)^2 = (\sigma_u t)^2 \qquad (5.4.11)$$

where $\sigma_u = \dfrac{U}{\sqrt{12}}$ is the standard deviation of fluid velocity over the cross section.

It is a general feature of pure advection that σ_x or the cloud size grows linearly with time, while gradient diffusion leads to $\sigma_x \propto \sqrt{t}$, as we shall see in Section 5.4.4.

5.4.4 Gradient diffusion and the diffusion equation

In Example 5.4.1 it was assumed that the only velocities present were those given by the macroscopic Poiseulle velocity distribution (5.4.9). Hence, it was possible for the initially dyed fluid body to remain sharply identifiable, having the concentration (5.4.8) while the concentration remained zero everywhere else. In reality this is never exactly true. Even in laminar flow the solute and water molecules undergo Brownian motions in all directions, which gradually modify the solute distribution. The paraboloid shells of dye in Figure 5.4.2 will become fussy and the overall distribution characteristics will be affected.

In still water (or water that moves with uniform velocity) a drop of solute will, due to the Brownian motions, spread in accordance with

$$\sigma_x^2 = 2Dt \qquad (5.4.12)$$

where D is called the molecular diffusivity. Typical values of molecular solute diffusitivities in water are 10^{-10} to $5 \times 10^{-9} \mathrm{m}^2/\mathrm{s}$. The diffusivity of heat is much greater. It is similar to the kinematic viscosity v ($\sim10^{-6}\,\text{m}^2/\text{s}$), which can be regarded as the diffusivity of momentum in laminar flow.

Considered on scales much greater than the typical path length between Brownian collisions the flux $\vec{Q}$ of solute is towards lower concentrations at a rate given by Fick's law of gradient diffusion:

$$\vec{Q} = -D\,\overrightarrow{\text{grad}C} \qquad (5.4.13)$$

Example 5.4.2: Flux due to swapping

The Brownian diffusion process is analogous to the mixing that occurs when say, red and green marbles are swapped randomly between two jars without changing the total number in each jar as in Figure 5.4.3.

If the marbles are sorted with one colour in each jar at the start, random swapping will initially result in a net transport of greens into the "red" jar and vice versa. Eventually, equilibrium is established with approximately equal numbers of each colour in each jar.

Figure 5.4.3:
The flux of marbles of one colour from one jar to another by random swapping is analogous to gradient diffusion. If one jar contains predominantly one colour, random swapping leads to a flux of this colour into the other jar.

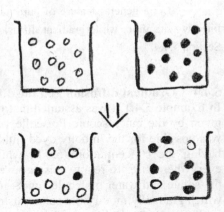

At each stage the flux of each colour can be estimated in analogy with the gradient diffusion formula (5.4.13), the concentration difference in between jars for each colour plays the role of $\overrightarrow{\text{grad}C}$ for that colour, and the swapping rate is analogous to the diffusivity.

☺

Divergence of the $\vec{Q}$-field will cause local concentration changes in accordance with the conservation of solute mass:

$$\frac{\partial C}{\partial t} = -\nabla \cdot \vec{Q} \qquad (5.4.14)$$

which, if the flux given by (5.4.13), with constant D, becomes

$$\frac{\partial C}{\partial t} = D \Delta C = D \left(\frac{\partial^2 C}{\partial x^2} + \frac{\partial^2 C}{\partial y^2} + \frac{\partial^2 C}{\partial z^2} \right) \qquad (5.4.15)$$

Equations (5.4.13) and (5.4.15) hold exactly for heat transport in homogeneous solids and a large number of analytical solutions for different geometries have therefore been developed, see e g, Crank (1956) or Carslaw and Jaeger (1959). A few of these solutions are of relevance as qualitative guides for dispersion of solutes in flowing water or air.

Firstly, for the one-dimensional case where diffusive fluxes are in the x-direction only, and the dye is released at x_0 at time t_0, we have

$$C(x,t) = \frac{Vo}{2\sqrt{\pi D(t-t_o)}} e^{-\frac{(x-x_o)^2}{4D(t-t_o)}} \qquad (5.4.16)$$

That is, a normal distribution centred at x_0 with variance

$$\sigma^2(t) = 2D(t-t_0) \qquad (5.4.17)$$

Figure 5.4.4 shows the development of the spreading distribution with time.

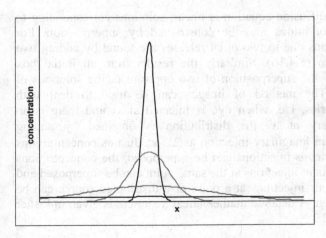

Figure 5.4.4: Gaussian distributions corresponding to (5.4.16) at successive times (0.1, 1, 10, 100) corresponding to gradient diffusion from an instantaneous point injection.

Secondly, if the concentration is a step of height C_0 at $x=x_0$ ($C = C_0$ for $x<x_0$, $C = 0$ for $x>x_0$) up to $t=t_0$ after which mixing is allowed in accordance with (5.4.13), the concentrations will be given by

$$C(x,t) = \frac{Co}{2}\left[1 - erf\left(\frac{x - x_o}{\sqrt{4D(t - t_o)}}\right)\right] \quad \text{for } t > t_0 \qquad (5.4.18)$$

see Figure 5.4.5, where the error function is defined by $erf(z) = \frac{2}{\sqrt{\pi}}\int_o^z e^{-\varsigma^2} d\varsigma$.

Figure 5.4.5:
Successive concentration distributions corresponding to the error-function solution (5.4.18) at $t-t_0=$ 01, 1, 10, 100.

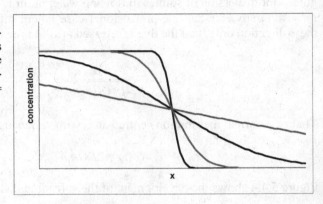

Because the diffusion equation is linear, solutions corresponding to more complicated conditions may be constructed by superposition. For example, concentrations due to two point releases are found by adding two solutions of the form (5.4.16). Similarly the results from an initial box shaped distribution is the superposition of two opposite facing solutions of the form (5.4.18). The method of images can be used to deal with impermeable boundaries, i e, when dye is injected at x_0 and there is an impermeable boundary at X, the distribution is obtained by adding concentrations from an imaginary injection at $2X-x_0$. Just as concentrations from several simultaneous injections can be superposed, the concentrations due to several subsequent injections at the same point can be superposed and the result of continuous injections at a point or a distributed source can be obtained by integration. A detailed mathematical treatment is given in Fisher et al. (1979) Section 2.3.

5.4.5 Advection diffusion

The equations (5.4.13) and (5.4.15) are valid if the fluid is at rest, or in a frame of reference which follows the fluid. If the fluid is moving with a steady, uniform velocity $\vec{u}$, we must add advective fluxes everywhere:

$$\vec{Q} = -D\ \overrightarrow{grad\ C} + \vec{u}\,C \tag{5.4.19}$$

Consequently, the conservation of mass equation (5.4.14) becomes (after assuming incompressible fluid: div $\vec{u} = 0$)

$$\frac{\partial C}{\partial t} = D\left(\frac{\partial^2 C}{\partial x^2} + \frac{\partial^2 C}{\partial y^2} + \frac{\partial^2 C}{\partial z^2}\right) - u\frac{\partial C}{\partial x} - v\frac{\partial C}{\partial y} - w\frac{\partial C}{\partial z} \tag{5.4.20}$$

This equation is called the *advection diffusion equation.*

Mathematically the superposition of a steady uniform fluid velocity does not add a great deal of difficulty, all that happens is that the whole picture drifts downstream with the steady uniform velocity $\vec{u}$. Hence, the only modification required for the standard one-dimensional solutions (5.4.16) and (5.4.18) is to replace x_o by $x_o + ut$.

For the important special case of continuous release like Figure 5.4.1 of $\dot{V}$ [m³/s] at a point (x_o, y_o, z_o), corresponding to a sewerage pipe issuing into a large river, a useful approximate expression for the downstream concentrations, which is illustrated by Figure 5.4.6, is

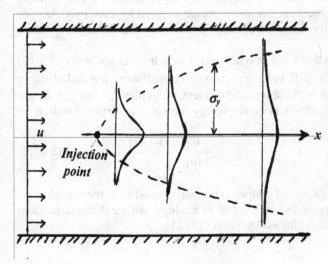

Figure 5.4.6: Consecutive downstream concentration distributions from a steady point injection given by (5.4.21). The dotted lines indicate the $x^{1/2}$ increase of the cross-stream spread. Effects from the presence of the river-banks are neglected. These can be handled using the method of images.

$$C(x,y,z) = \frac{\dot{V}}{4\pi D \frac{x-x_o}{u}} e^{-\frac{(y-y_o)^2+(z-z_o)^2}{4D(x-x_o)/u}} \quad \text{for } x > x_0+2D/u \qquad (5.4.21)$$

This solution assumes the same diffusivity in the y- and z-directions and neglects diffusion in the x-direction. The latter is reasonable from some distance downstream. Fisher et al. (1979) give the limit $x-x_o = 2D/u$.

5.4.6   The random walk analogy of turbulence and turbulent mixing
In a turbulent flow the chaotic turbulent motion will have an effect which, when viewed on a large enough scale, is similar to that of the Brownian motions though usually orders of magnitude are stronger. The turbulent swapping of fluid between layers is also analogous to the swapping process discussed in Example 5.4.2.

These various swapping processes can be modelled as random walk processes. That is, processes where individual particles or fluid elements move in random steps, which may be more or less correlated. Motion with constant velocity corresponds to maximum correlation (no change). Random walk governed by the roll of a dice in each step has zero correlation between successive velocities (or step lengths).

If each particle in a cloud rolls a dice to determine the length (in metres) of its next step, this cloud will move with average velocity 3.5m/s and spread at the rate $\sigma^2(t) = 2\times(35/12)\times t$ (3.5 and 35/12 being respectively the mean and variance of the "dice probability distribution"). It also follows from the central limit theorem that the distance covered by each particle after many steps is normally distributed, corresponding to the diffusion solution (5.4.16).

The $\sqrt{t}$ growth of the spread ($\sigma(t) \propto \sqrt{t}$) in analogy with (5.4.16) will however only result if velocities in subsequent steps are independent (as with the dice). In turbulence, subsequent velocities at spacing δ_t are correlated to an extent which, is quantified by the autocorrelation function

$$\rho_{uu}(\delta_t) = \frac{E\{[u(t-\delta_t)-\overline{u}][u(t)-\overline{u}]\}}{Var\{u\}} \qquad (5.4.22)$$

where $E\{\}$ denotes expected value. The way in which, the continuous movements in turbulence can be treated in analogy with a discrete random walk process, was first explained by Taylor (1921).

Measured autocorrelation functions from turbulence are usually similar to an exponential $\rho_{uu}(\delta_t) \approx e^{-\delta_t/T_L}$ except in the limit $\delta_t \to 0$, where it must have zero derivative, see Figure 5.4.7. T_L is called the Lagrangian integral time scale which is related to the dominant eddy size (the turbulent length scale) L and the typical turbulent velocity u' by $T_L \approx L/u'$.

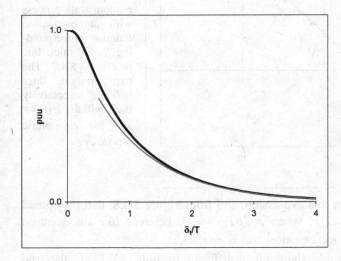

Figure 5.4.7: Qualitatively typical velocity auto-correlation function for turbulence. The thin line is $e^{-\delta_t/T_L}$.

Figure 5.4.8 shows the development of the spread $\sigma(t)$ for a cloud of particles, who each perform random walk in accordance with the autocorrelation function in Figure 5.4.7. G I Taylor showed that the spread of a cloud of particles in homogeneous, isotropic turbulence with velocity variance u'^2 and autocorrelation ρ_{uu} is given by

$$\sigma^2(t) = 2u'^2 \int_0^t\int_0^\theta \rho_{uu}(\tau)\,d\tau\,d\theta \qquad (5.4.23)$$

which, with the simple exponential autocorrelation function $\rho_{uu}(\tau) \approx e^{-\tau/T_L}$, gives

$$\sigma^2(t) = 2u'^2T_L^2\left(\frac{t}{T_L}+e^{-t/T_L}-1\right) = u'^2T_L^2\left((\frac{t}{T_L})^2-\frac{1}{3}(\frac{t}{T_L})^3+...\right) \qquad (5.4.24)$$

195

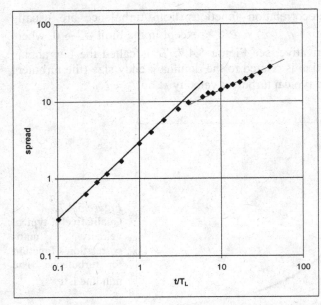

Figure 5.4.8:
Development of the size or spread of a cloud from a point source by a simple random walk process with autocorrelation function corresponding to the thick line in Figure 5.4.7. The two straight lines indicate respectively the initial $\sigma(t) \propto t$ and the asymptotic $\sigma(t) \propto \sqrt{t}$.

We see from this expression, and from Figure 5.4.8, that the spread eventually, for $t >> T_L$, when $\rho_{uu}(t/T_L) \rightarrow 0$, behaves like the solution (5.4.16) to the diffusion equation, that is $\sigma(t) \propto \sqrt{t}$.

Correspondingly, the gradient diffusion formula (5.4.13) is also true in this range, i e, the flux is proportional to the concentration gradient for $t >> T_L$. The line to the right is given by

$$\sigma^2(t) = 2u'^2 T_L t \qquad (5.4.25)$$

Hence, it is seen from comparison with (5.4.17) that the diffusivity corresponding to the turbulent velocity scale u' and the Lagrangian time scale T_L is

$$D = u'^2 T \text{ for } t >> T_L \qquad (5.4.26)$$

Alternatively this may be written in terms of the turbulent length scale (the mixing length) $L = u'T_L$:

$$D = u'L \text{ for } t >> T_L \qquad (5.4.27)$$

196

Initially, for $t \ll T_L$, when the velocities are highly correlated, Equation (5.4.24) shows that the spread grows linearly with time. The line on the left in Figure 5.4.8 is given by $\sigma(t) = u't$, which is seen to be analogous to the advective result (5.4.7). In this initial stage, the flux is not related to the concentration gradient, i e, Equation (5.4.13) does not apply. In other words, at this initial stage the turbulent diffusion process is not Fickian.

5.4.7 The mixing length approach to turbulent diffusion

Prandtl (1925) introduced the concept of a mixing length l_m which can be thought of as the typical diameter of the turbulent eddies which dominate the mixing process. By considering a swapping process analogous to that in Example 5.4.2, but with swaps between different layers l_m apart, rather than between jars, we can construct a useful quantitative model for turbulent diffusion. Consider the simplified scenario in Figure 5.4.9.

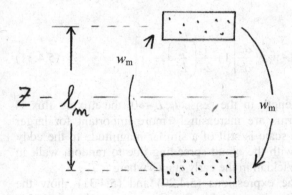

Figure 5.4.9:
The down-gradient flux of, for example sediment, due to turbulent mixing can be understood in terms of swapping fluid-sediment-mix parcels over a vertical distance l_m, the mixing length.

The sediment concentration in the lower parcel is $c(z-l_m/2)$ and that in the upper parcel $c(z+l_m/2)$. If the parcels travel vertically with equal and opposite velocities $\pm w_m$, the resulting flux density is

$$q_m = w_m[c(z-l_m/2) - c(z+l_m/2)] \qquad (5.4.28)$$

and by Taylor expansion of $c(z \pm l_m/2)$

$$q_m = -w_m l_m \left[\frac{dc}{dz} + \frac{l_m^2}{24} \frac{d^3 c}{dz^3} + \dots \right]$$

$$= -w_m l_m \frac{dc}{dz} \left[1 + \frac{l_m^2}{24} \frac{\frac{d^3 c}{dz^3}}{\frac{dc}{dz}} + \dots \right] \qquad (5.4.29)$$

We recognize the term in front of the last square bracket as being the usual Fickian or "gradient diffusion" flux density corresponding to the diffusivity $D = l_m w_m$. The nature of the contents of the bracket is perhaps best illustrated by considering the special case of an exponential concentration distribution with vertical scale L_c, i e,

$$c(z) = C_o e^{-z/L_c} \qquad (5.4.30)$$

In this case we get

$$q_m = -w_m l_m \frac{dc}{dz} \left[1 + \frac{1}{24} \left(\frac{l_m}{L_c} \right)^2 + \dots \right] \qquad (5.4.31)$$

that is, for small mixing length in the sense $l_m/L_c \to 0$, the mixing flux is Fickian. The additional terms are increasingly more important for larger l_m/L_c, i e, when the overall scale is still of a similar magnitude to the eddy size l_m. Note the analogy with the cloud spreading due to random walk in Figure 5.4.8, which is non-Fickian in the early stages where $t < T_L$.

In other words, the expressions (5.4.29) and (5.4.31) show the limitations of the Fickian diffusion model in Eulerian terms as Taylor (1921) showed it in Lagrangian terms.

5.4.8 Apparent Fickian diffusivities

Even when a particular mixing process is known to be not strictly Fickian, the Fickian model is often adopted as a convenient approximation. A nominal or apparent Fickian diffusivity K_{Fick} may then be defined, either in terms of observed mixing fluxes and gradients

$$K_{Fick} = -\frac{q_m}{gradient} \qquad (5.4.32)$$

in accordance with Fick's law, Equation (5.4.13), or in terms of observed rates of spreading

$$K_{\text{Fick}} = \frac{1}{2} \frac{d}{dt} \sigma^2(t) \qquad (5.4.33)$$

in accordance with (5.4.17).

These nominal Fickian diffusivities, though commonly used in engineering praxis, often show very peculiar behaviours: different species, e g, mass, momentum, heat, salt, sediment, bubbles, temperature, ... may have very different K_{Fick}-values in a given flow. This variability is generally referred to as *differential diffusion*. It can, at least in some cases, be understood in terms of (5.4.29) in the form

$$q_m = -w_m l_m \frac{dc}{dz} \left[1 + O\left(\frac{l_m}{L}\right)^2 \right] \qquad (5.4.34)$$

which emerges when the length scale L is defined as $(c'/c''')^{1/2}$ (Some consideration of the interaction of the particular species with vortices may also be needed for a detailed explanation).

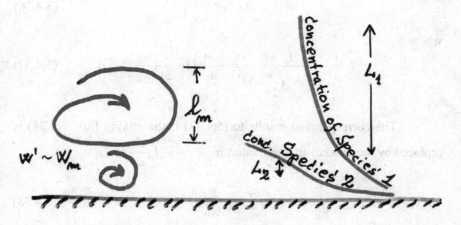

Figure 5.4.10: Different species with different vertical scales will, in accordance with (5.4.34), display different apparent Fickian diffusivities, $K_{\text{Fick}} = -q_m/(dc/dz)$ in a given flow. So, while l_m and w_m are entirely determined by the flow, K_{Fick} is not.

The most peculiar case of *differential diffusion* is perhaps that of different sediment sizes with different settling velocities, w_s, suspended in the same flow. That is, measured concentration distributions $c(z)$, e g, Anderson (1942), Coleman (1970), Graf and Cellino (2002) for rivers and Nielsen (1983) under waves, consistently show K_{Fick} to be an increasing function of $w_s/u*$. This means that, if the mixing process were Fickian, $w_m^2 T_L$ [cf Equation (5.4.26)] must be greater for larger, more inert particles, than for small ones. No satisfactory explanation for such a strange state affairs has ever been found. However, the finite mixing length model above can explain the observations at least partly.

What happens is that, particles with greater settling velocity will be more concentrated close to the bed, i e, they have a smaller vertical concentration scale L_c and the second term in the flux formula (5.4.34) makes a greater contribution. Quantitatively it works as follows: If the situation is steady, there is a balance between the mixing flux (5.4.28) and the settling flux $w_s c(z)$:

$$w_m[c(z-l_m/2) - c(z+l_m/2)] - w_s c(z) = 0 \qquad (5.4.35)$$

For the simple case of homogeneous turbulence, where both w_m and l_m are constant this has the solution

$$c(z) = c(z_o)e^{-(z-z_o)/L} \qquad (5.4.36)$$

where

$$L = \frac{l_m}{2\sinh^{-1}(\dfrac{w_s}{2w_m})} = \frac{l_m w_m}{w_s}[1+\frac{1}{24}(\frac{w_s}{w_m})^2 - ...] \qquad (5.4.37)$$

The corresponding results to (5.4.37) if the mixing flux (5.4.28) is replaced by the Fickian approximation $q_{Fick} = -w_m l_m \dfrac{dc}{dz}$ is

$$L_{Fick} = \frac{l_m w_m}{w_s} \qquad (5.4.38)$$

i e, the true vertical scale is greater than the Fickian approximation and more so for coarser sediment (for greater w_s). This increase in L with w_s is synonymous with the apparent Fickian diffusivities increasing with w_s

because the apparent Fickian diffusivity is $w_s L$. Together with the result (5.4.37) this gives

$$K_{\text{Fick}} = w_m l_m \left[1 + \frac{1}{24} (\frac{w_s}{w_m})^2 + ... \right] \qquad (5.4.39)$$

a result which is in good qualitative agreement with the differential diffusion observations of different sediment sizes in the same flow.

5.4.9 Fickian approximations to turbulent diffusion

The considerations above indicate that analogous equations to (5.4.12) and (5.4.13) will hold for turbulent diffusion at sufficiently large time- or length-scales ($t > T_L$, $|\vec{r}| > u'T_L$). However, the turbulent mechanisms are rarely isotropic like the Brownian motion. That is, there is generally a different turbulent diffusivity in each direction and hence, there is usually not a simple turbulent equivalent to (5.4.13).

Instead, one may define a diffusivity tensor (matrix) K_{ij} by:

$$\vec{Q} = - \begin{bmatrix} K_{xx} & K_{xy} & K_{xz} \\ K_{yx} & K_{yy} & K_{yz} \\ K_{zx} & K_{zy} & K_{zz} \end{bmatrix} \begin{bmatrix} \dfrac{\partial C}{\partial x} \\ \dfrac{\partial C}{\partial y} \\ \dfrac{\partial C}{\partial z} \end{bmatrix} \qquad (5.4.40)$$

where the non-diagonal terms can usually be neglected if the main direction of flow is in the direction of one of the coordinate axes, i e

$$\vec{Q} \approx - \begin{bmatrix} K_{xx} \dfrac{\partial C}{\partial x} \\ K_{yy} \dfrac{\partial C}{\partial y} \\ K_{zz} \dfrac{\partial C}{\partial z} \end{bmatrix} \qquad (5.4.41)$$

For constant diffusivities analytical solutions can still be obtained in analogy with the solutions to (5.4.13), e g, the spread from an instantaneous point injection at $(x_o, y_o, z_o.0)$ in three dimensions with diffusivity tensor

$$K_{ij} = \begin{bmatrix} K_{xx} & 0 & 0 \\ 0 & K_{yy} & 0 \\ 0 & 0 & K_{zz} \end{bmatrix} \qquad (5.4.42)$$

is given by

$$C(x,y,z,t) = \frac{1}{\sqrt{4\pi K_{xx}t}\sqrt{4\pi K_{yy}t}\sqrt{4\pi K_{zz}t}} e^{-\left(\frac{(x-x_o)^2}{4K_{xx}t} + \frac{(y-y_o)^2}{4K_{yy}t} + \frac{(z-z_o)^2}{4K_{zz}t}\right)} \qquad (5.4.43)$$

In reality however, the diffusion tensor may well be a function of both time and position. For example, K_{zz} is related to the vertical diffusivity of horizontal momentum: the eddy viscosity v_t, which for a steady uniform flow with a logarithmic velocity distribution $u(z) = \dfrac{u_*}{\kappa}\ln(\dfrac{z}{k_s/30})$ is a parabolic function of z:

$$v_t = \kappa z u_*(1-\frac{z}{h}) \qquad (5.4.44)$$

cf Section 6.2.2. Simply depth averaging (5.4.44) to get a representative D_{zz} gives

$$K_{zz} \approx <v_t> = 0.067\, h\, u_* \qquad (5.4.45)$$

Fisher et al. (1979) present data for K_{yy} in both prismatic and irregular streams. For straight, rectangular channels they may be summarised as

$$\frac{K_{yy}}{hu_*} = 0.16 \pm 0.07 \qquad (5.4.46)$$

while the values are larger for irregular channels:

$$\frac{K_{yy}}{hu_*} = 0.75 \pm 0.81 \qquad (5.4.47)$$

(while the standard deviation is greater than the mean, negative values were of course not encountered. The range was 0.30 to 3.4).

Fisher et al.'s data for measured longitudinal dispersion in natural streams correspond to

$$\frac{K_{xx}}{hu_*} = 524 \pm 1491 \tag{5.4.48}$$

The large standard deviation is mainly due to a single observation at 7500 while the rest of the values lie between 20 and 500 with a modal value close to 200.

These values should only be taken as indicators of general magnitude and variability. In a concrete modelling situation, one should place particular emphasis on data from situations similar to that which is being modelled, and first read the more comprehensive texts on this very complex issue, e g, Fisher et al. (1979) and Rutherford (1994).

5.4.10 Dispersion due to combined shear, advection and diffusion

The very great difference in magnitude between K_{xx} and K_{yy} is due to the fact that the streamwise spreading is a combination of shear flow advection, exemplified in Figure 5.4.2, and diffusion in the direction perpendicular to the mean flow.

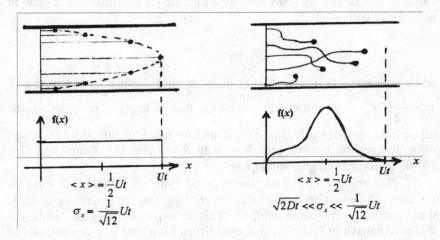

Figure 5.4.11: Streamwise dispersion of solute released uniformly across the pipe at $x=0$ and $t=0$ in a circular pipe with parabolic $\bar{u}(r)$. *Left*: Without lateral diffusion. *Right*: With lateral diffusion. In both cases, the centre of mass moves with speed $<u> = U/2$. Without diffusion, the density function $f(x)$ is rectangular at all times. Lateral diffusion reduces σ_x.

203

The interaction between the two mechanisms was investigated by G I Taylor (1953, 1954) for laminar and turbulent pipe flows. Taylor's pioneering insight was that, after some considerable time (or distance from the injection point), a cloud of solute will spread in the flow direction in accordance with Fickian diffusion with diffusivity K_∞.

For laminar flow in a circular pipe, he derived the following approximate but quite accurate result:

$$K_\infty = \frac{1}{192}\frac{R^2 U^2}{D} \text{ for } x > 0.2\frac{R^2 U}{D} \qquad (5.4.49)$$

where D is the molecular diffusivity, R is the pipe radius and U the velocity at the centreline. It is truly remarkable that K_∞ is inversely proportional to D. This is however in accordance with the fact that the wandering of particles across the pipe due to diffusion, reduces the spread of time mean velocities for individual particles, see Figure 5.4.11. Hence the streamwise dispersion rate is reduced by D compared with purely advective dispersion. With lateral diffusion, but still with a parabolic $\bar{u}(r)$, f(x) is initially rectangular, but tends towards a "Gaussian shape".

The asymptotic Fickian behaviour means that, the streamwise variance will ultimately grow linearly with time (cf Equation 5.4.12) and in accordance with

$$\frac{d\sigma_x^2}{dt} = 2K_\infty \qquad (5.4.50)$$

while pure shear flow advection as discussed in Example 5.4.1, would give $\sigma_x^2 = \frac{1}{12}U^2 t^2$, i e, growth as t^2 rather than linearly. Nevertheless the asymptotic apparent diffusivity K_∞ is usually much greater than D. As an example consider laminar pipe flow with $R = 3$mm, $U = 2$cm/s and $D = 10^{-9}$m^2/s corresponding to salt in water.

Equation (5.4.33) then gives $K_\infty = 0.019$m^2/s $\sim 19000000 D$, valid for $x > 36$m. The initiation zone $x < 36$m is very long but, the magnitude of K_∞ is truly enormous compared with D. This example shows that, while the diffusive behaviour (5.4.50) is not established for a long way downstream of the release, the apparent streamwise diffusivity must quickly become much greater than D. The natural stream equivalent is that K_{xx} very quickly becomes greater than $u_* h$.

The asymptotic behaviour described by (5.4.50) is independent of the details of the release conditions because it corresponds to the solute having been well mixed across the pipe by lateral diffusion. The initial behaviour of σ_x will, on the other hand, naturally depend on the details of the release conditions. If the dye is released as a uniform sheet across the pipe at $x=0$ and $t=0$, the streamwise dispersion will initially be as in the purely advective dispersion, discussed in Example 5.4.1, that is, with initial uniformity across the pipe lateral diffusion has no effect initially. The scenario is then as in Figure 5.4.8.

Analogous results to (5.4.49) hold for various turbulent flows. For turbulent pipe flow the general magnitude of the transverse diffusivity ($\approx$ the eddy viscosity) is the pipe radius times the friction velocity: $K_{turbulent} \approx Ru_*$. Hence, as the centreline velocity U is also proportional to u_*, the turbulent pipe flow analogy to (5.4.49) is of the form

$$K_\infty = 10.1 Ru_* \text{ for } x > 0.8R\frac{U}{u_*} \tag{5.4.51}$$

The numerical coefficient, which depends on the velocity distribution was obtained by Taylor (1954). For uniform turbulent flow of depth h and logarithmic velocity distribution down an infinitely wide slope the analogous result is

$$K_\infty = 5.93 hu_* \text{ for } x > 0.5h\frac{U}{u_*} \tag{5.4.52}$$

The stages of dispersion of a cloud of solute into a stream can now be summarised as in Figure 5.4.12:
- Initially, the solute motion is closely correlated with its release velocity and the cloud size (σ_y, σ_z) grows linearly with time. This is the case while the cloud size is smaller than the dominant eddy size or while $|\vec{r} - \vec{r}_o|/u' < T_L$. This stage corresponds to the line on the left of Figure 5.4.8.
- For $|\vec{r} - \vec{r}_o|/u' > T_L$ the diffusion equations with respective diffusivities K_{yy} and K_{zz} hold in the y and z directions and the solute distributions in these directions tend towards normal distributions possibly modified by influence from the sides and bottom of the channel.
- For $|\vec{r} - \vec{r}_o|/u' > 0.4W^2/K_{yy}$ the vertical and lateral distributions reach an equilibrium with $\dfrac{\partial <C>_x}{\partial x}$ and the streamwise dispersion rate is as in Fickian diffusion with diffusivity K_∞.

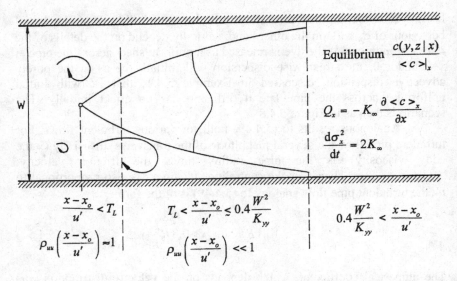

$$\text{Equilibrium} \quad \frac{c(y,z\,|\,x)}{<c>|_x}$$

$$Q_x = -K_\infty \frac{\partial <c>_x}{\partial x}$$

$$\frac{d\sigma_x^2}{dt} = 2K_\infty$$

$$\frac{x-x_o}{u'} < T_L \qquad T_L < \frac{x-x_o}{u'} \le 0.4\frac{W^2}{K_{yy}} \qquad 0.4\frac{W^2}{K_{yy}} < \frac{x-x_o}{u'}$$

$$\rho_{uu}\left(\frac{x-x_o}{u'}\right) \approx 1 \qquad \rho_{uu}\left(\frac{x-x_o}{u'}\right) \ll 1$$

Figure 5.4.12: The different stages in the dispersion process from a source in a turbulent river.

5.4.11 Dispersion in oscillatory shear flow

The powerful combination of transverse diffusion and shear flow advection also leads to very large stream wise dispersion rates in oscillatory flows like tidal flows in rivers and in the airflow induced by breathing. In fact, it is because of these very large dispersion rates that humans can get sufficient oxygen to the ends of the airways while actually only emptying the lungs 1/3 in normal breathing.

However, just as the initial ranges before the validity of (5.4.49) in steady flows are quite long, it requires a reasonably long oscillation period for the streamwise dispersion in oscillatory flows to take the gradient diffusion form. Fisher et al. (1979) give an analysis of the phenomenon, which leads to

$$K_\infty = \frac{1}{240}\frac{h^2U^2}{D} \approx \frac{1}{240}\frac{h^2U^2}{0.067u_*h} \approx 0.06\frac{hU^2}{u_*} \quad \text{for } T > \frac{h^2}{D} \approx 200\frac{h}{u_*}$$

(5.4.53)

where D is the diffusivity perpendicular to the main flow and U is the velocity amplitude of the flow.

6 Sediment transport mechanisms

6.1   INTRODUCTION

For an introductory overview of the sediment transport modes on a beach consider Figure 6.1.1.

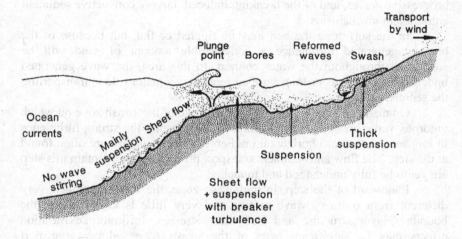

Figure 6.1.1: Sediment transport modes on a beach, a simplified 2D view. 3D features like rip currents and longshore currents will increase the vertical scale of entrainment and transport sand. On fine grained beaches the step at the bottom of the swash zone may be absent.

In deep water, the bed is unaffected by waves and consequently any sediment transport will be due to tidal or other ocean currents.

In somewhat shallower depths ($h<25$m) the wave motion will often dominate flow near the bed and the sediment transport. As an indication of this, the bed will often be covered by sharp crested, shore-parallel ripples (vortex ripples), over which most of the sediment is transported in suspension.

Close to the breakpoint the wave induced water motion at the bed is often so vigorous that the vortex ripples disappear and the bed will have megaripples or be almost perfectly flat. In this area, the transport occurs essentially within a few millimetres of the bed level. The transport mode which is called sheet flow involves a mixture of bedload and suspended load.

Around the breakpoint, the sediment transport occurs throughout the depth because the breaking waves can generate strong vertical flows, bringing large amounts of sediment towards the surface. This sediment, being suspended for several wave periods, is then transported by the currents including wave generated circulation. This area still holds many challenges including description of the water motion, which is not at all like that of progressive waves, and of the breaking induced, largely convective sediment entrainment mechanisms.

In the surf zone the bed may be rippled or flat, but because of the breaker generated turbulence a considerable amount of sand will be suspended throughout the water column. In this area, the wave generated undertow, rip currents or longshore currents play a major role in transporting the sediment.

On many beaches a step exists at the base of the swash zone on which vigorous wave breaking occurs. Surprisingly, despite the strong turbulence in this area, very sharp horizontal gradients in sediment size are often found at the step. The flow and sediment transport processes that maintain this step are yet to be fully understood and modelled.

Landward of the step, in the swash zone, the water motion is very different from ordinary wave motion and very little is known about the boundary layer structure and bed shear stresses. Infiltration/exfiltration effects may be significant parts of the swash zone sediment transport mechanisms. While the sand surface in the swash zone is usually flat, sandwaves similar to antidunes in supercritical steady flow sometimes develop during long lasting backwash events, see Figure 2.3.8. The sediment transport in the swash zone is very significant in the shore parallel as well as the shore normal direction.

From the top of the swash zone the sand may be moved on by onshore winds.

6.2 BED SHEAR STRESSES

6.2.1 Introduction

Sediment particles are entrained by the flow via the bed shear stresses $\tau_b(t)$. In general, both waves and currents contribute to the bed shear stress.

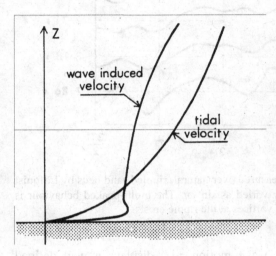

Figure 6.2.1:
Rapidly accelerated flows induced by waves have thinner boundary layers than longer period flows like tides. The boundary layer thickness δ is proportional to the square root of the wave period, $\delta \propto (v_t T)^{1/2}$. Consequently, since $\tau_b \propto v_t u_\infty / \delta$, wave motions generate greater bed shear stresses for a given free stream velocity (u_∞) magnitude.

However, since the wave boundary layers are thinner, the waves tend to generate greater bed shear stresses for a given velocity magnitude, see Figure 6.2.1.

Experimental evidence indicates that the wave boundary layer structure and the wave shear stresses are practically unaffected by the presence of currents, cf Nielsen (1992) Figure 1.5.8 and Simons et al. (2000). Hence, for practical purposes, the sediment entrainment can usually be contributed to the wave-generated stresses and these can be calculated as if the current was not there.

Unsteady bed shear stresses are very difficult to measure over movable sand beds. Consequently, our knowledge about them is still limited to two experimental studies: Carstens et al. (1969) and Löfquist (1986). Carstens et al. only measured the time-averaged energy dissipation and hence they provide no information about the shape of $\tau_b(t)$. Löfquist did measure time series of $\tau_b(t)$, but only on rippled beds where the behaviour of $\tau_b(t)$ is quite complicated because of vortex formation, see Figure 6.2.2.

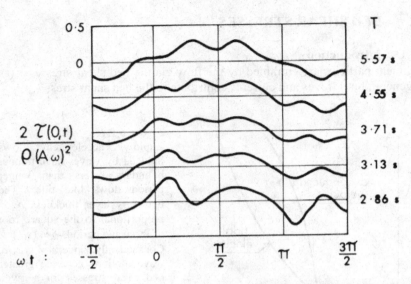

Figure 6.2.2: Bed shear stress measured over natural, rippled sand beds by Löfquist (1986). The free stream velocity varied as sin ωt. The multi-peaked behaviour is due to the rhythmic formation of vortices at the ripple crests.

For simple harmonic wave motion $\tau_b(t)$ displays a well defined phase lead compared with the free stream velocity. For laminar flow, this phase lead is 45°, cf Nielsen (1992) p. 21. For turbulent flows it is less.

For sediment transport purposes, the total bed shear stress on a bed with bedforms, e g ripples, is often split into the form drag $\tau''(t)$ and the skin friction $\tau'(t)$.

$$\tau(t) = \tau'(t) + \tau''(t) \qquad (6.2.1)$$

This is because the form drag, which is related to the front-and-back pressure differences on the bed forms, is considered irrelevant to the movement of sediment particles. The sediment is moved by the skin friction.

6.2.2 The log law for steady flow

In a steady uniform flow with depth h and surface slope S, the bed shear stress must balance the downhill pull by gravity. This gives

$$\overline{\tau_b} = \rho g h S \qquad (6.2.2)$$

210

and correspondingly, the steady friction velocity is given by

$$\overline{u_*} = \sqrt{\tau_b / \rho} = \sqrt{ghS} \qquad (6.2.3)$$

A reasonable fit to steady flow velocity profiles and a simple, yet useful theoretical model is provided by the logarithmic velocity profile also known as "the law of the wall"

$$\overline{u}(z) = \frac{\overline{u_*}}{\kappa}\ln\frac{z}{z_o} \approx \frac{\overline{u_*}}{\kappa}\ln\frac{z}{r/30} \qquad (6.2.4)$$

where κ is von Karman's constant and r is the hydraulic roughness (Nikuradse roughness) of the bed.

The *law of the wall* was developed by Theodore von Karman in the 1920s but its validity or otherwise is perhaps most succinctly summarized by the derivation given by Landau and Lifshitz (1987). That is: Assume as in Figure 6.2.3 that a layer exists where the velocity gradient is determined completely by the bed shear stress, the fluid density and the elevation.

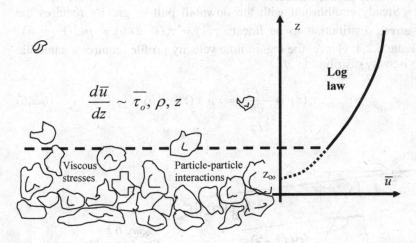

Figure 6.2.3: In a steady, uniform flow, an important layer exists where the shape of the velocity profile is governed by the bed shear stress and the fluid density. In this layer the velocity distribution must be logarithmic. In the layer below, viscous stresses and/or intergranular forces from moving sediment will also influence $\overline{u}(z)$.

If indeed $\dfrac{d\bar{u}}{dz}, \bar{\tau_o}, \rho, z$ are the only relevant physical variables, it follows from dimensional analysis that only one dimensionless combination is possible and that it must be a constant. One way of expressing this is

$$\frac{z\dfrac{d\bar{u}}{dz}}{\sqrt{\bar{\tau_o}/\rho}} = \text{constant} \qquad (6.2.5)$$

where the constant is written as $1/\kappa$, and $\kappa \approx 0.4$ is called von Karman's constant.

Integrating and introducing the friction velocity $\bar{u}_* = \sqrt{\bar{\tau_o}/\rho}$, this gives (6.2.4) where the constant of integration z_0 depends on the conditions in the *inner layer* below the log layer. The value $z_0 = r/30$ implied by (6.2.4) applies to fully rough turbulent conditions which are usually assumed in vigorous sediment transport scenarios, see Schlichting and Gersten (2000) for details.

Note that z_0 is also the level at which the downward extrapolation of the log velocity profile hits zero.

Steady equilibrium with the downhill pull of gravity requires the shear stress distribution to be linear: $\bar{\tau}(z) = \tau_b(1 - z/h) = \rho u_*^2(1 - z/h)$, see Figure 6.2.4. Hence, the logarithmic velocity profile requires a parabolic eddy viscosity distribution

$$\bar{v}_t(z) = \frac{\bar{\tau}/\rho}{\dfrac{\partial \bar{u}}{\partial z}} = \kappa \bar{u}_* z(1 - z/h) \qquad (6.2.6)$$

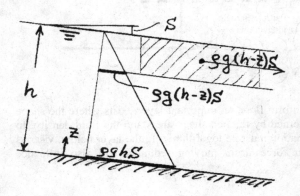

Figure 6.2.4:
Steady equilibrium with the downhill pull of gravity requires $\bar{\tau}(z)$ to vary linearly with the distance from the surface.

6.2.3　The wave friction factor

Since the time dependence of $\tau_b(t)$ is excessively complicated for rippled beds, Figure 6.2.2, $\tau_b(t)$ is often only modelled as far as its peak value $\hat{\tau}_b$ is concerned. Through the definition

$$\hat{\tau}_b = \frac{1}{2}\rho(A\omega)^2 f_w \qquad (6.2.7)$$

this then involves prediction of the wave friction factor f_w, which is a function of the relative bed roughness r/A and of the boundary layer Reynolds number $A^2\omega/v$. A is the water particle semi excursions just above the boundary layer, $= \alpha(-h)$ in the notation of Figure 1.2.1.

The behaviour of $f_w(r/A, A^2\omega/v)$ was investigated comprehensively by Riedel (1972) for fixed beds with variable sand grain roughness and his results are shown in Figure 6.2.5, which is analogous to the Moody diagram for pipe flow friction factors.

Note that $\tau_{b,\max}$ always occurs ahead of $u_{\infty,\max}$ even if the bed is not rippled as for the data in Figure 6.2.2. For laminar flow the phase lead is $45°$. For turbulent flows it is less, decreasing with increasing Reynolds number.

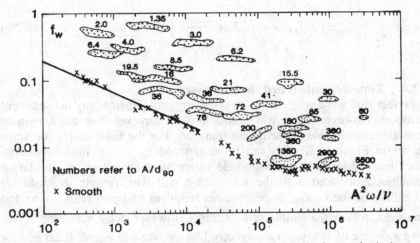

Figure 6.2.5: Wave friction factors for fixed flat sand beds, i e, a single layer of sand grains glued to the top of a shear plate, measured by Riedel (1972). The numbers refer to A/d_{90}. The solid line corresponds to the theoretical, smooth laminar flow result $f_w = \dfrac{2}{\sqrt{RE}} = \dfrac{2}{\sqrt{A^2\omega/v}} = $, cf Nielsen (1992) p. 21.

For most field applications, the Reynolds number dependence of f_w can be ignored, i e, the flow is considered "fully rough turbulent", and the friction factor is considered to be a function of the relative roughness only, e g,

$$f_w = \exp\left[5.5\left(\frac{r}{A}\right)^{0.2} - 6.3\right] \qquad (6.2.8)$$

which is within a factor 1.5 of the available fixed bed measurements for $0.0002 < r/A < 1$, cf Nielsen (1992) p. 26. Thus, calculation of the friction factor, for total bed friction, is not a problem, provided the hydraulic roughness r of the bed is known. Knowing what r is for movable sediment beds is, however, a very complex and largely unresolved issue. Further uncertainty arises with respect to the sediment transport because it is a function of the skin friction τ' only, rather than of the measurable total.

At the current state of the art, these complications are usually bypassed in sediment transport estimations, and the wave shear stresses and resulting sediment transport are dealt with in terms of the nominal *grain roughness Shields Parameter* $\theta_{2.5}$ and the corresponding friction factor $f_{2.5}$, which is determined from Equation (6.2.8) with $r = 2.5d_{50}$:

$$f_{2.5} = \exp\left[5.5\left(\frac{2.5d_{50}}{A}\right)^{0.2} - 6.3\right] \qquad (6.2.9)$$

6.2.4   Time dependent bed shear stresses

A perfect sine wave will, due to its symmetry, not generate any net sediment transport. However, other wave motions with zero net flow are known to generate considerable net sediment transport. For the finite-amplitude wave shapes in Figure 1.5.1 this can be understood by steady flow reasoning: Since the bed shear stress magnitude varies more or less as u^2, the larger velocities associated with the wave crest will win, despite their shorter duration. For such wave shapes, the net sediment transport rates, of not too fine sand, follow the simple rule $\overline{Q_s} \propto \overline{u_\infty^3}$. However, most surf zone waves have a degree of saw-tooth asymmetry like the wave in Figure 6.2.6 and for these waves the simple $\overline{Q_s} \propto \overline{u_\infty^3}$ rule does not work. Waves with saw-tooth asymmetry often generate significant shoreward sediment transport even though $\overline{u_\infty^3} = 0$ as well as $\overline{u}_\infty = 0$.

The reason for this landward transport is the strong asymmetry in the bed shear stress, which is indicated in Figure 6.2.6. When the flow is acceleration very abruptly, the boundary layer has had no time to grow, and the shear stresses become very large for a given value of the free stream velocity.

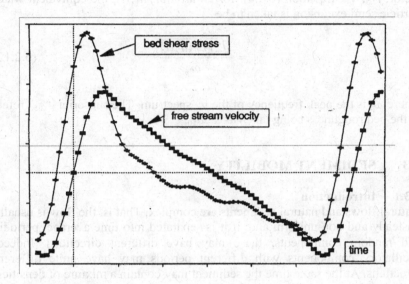

Figure 6.2.6: The thinner boundary layer during abrupt acceleration, leads to greater bed shear stresses for a given value of the free stream velocity.

While, as mentioned above, the time dependent shear stresses on flat, movable sand beds have not been directly measured, several kinds of indirect evidence (Nielsen 2006) support the formula

$$\tau_{2.5}(t) = \frac{1}{2}\rho f_{2.5}\left[\cos\varphi_\tau\, u_\infty(t) + \sin\varphi_\tau\, \frac{u_\infty(t+\delta_t) - u_\infty(t-\delta_t)}{2\omega_p\delta_t}\right]^2 \text{sign}[u_*(t)]$$

(6.2.10)

where δ_t is the time step of the free stream velocity $u_\infty(t)$ time series and φ_τ is the phase lead of the bed shear stress at the peak angular frequency ω_p. Its best-fit value for sheet flow, including swash zone, applications seems to be $\varphi_\tau \approx 51°$. The corresponding instantaneous friction velocity is

215

$$u_*(t) = \sqrt{\frac{1}{2}f_{2.5}}\left[\cos\varphi_\tau\, u_\infty(t) + \sin\varphi_\tau\, \frac{u_\infty(t+\delta_t) - u_\infty(t-\delta_t)}{2\omega_p\delta_t}\right] \quad (6.2.11)$$

For the purpose of calculating the grain roughness wave friction factor, $f_{2.5}$, via Equation (6.2.8), for an arbitrary $u_\infty(t)$, the equivalent water particle semi excursion is taken to be

$$A = \frac{\sqrt{2Var\{u_\infty(t)\}}}{\omega_p} \quad (6.2.12)$$

where ω_p is the peak frequency of the u_∞ spectrum. The subscript "$_{2.5}$" refers to the bed roughness being taken as $2.5d_{50}$.

6.3   SEDIMENT MOBILITY

6.3.1   Introduction

Natural flows and natural sediments are complex. That is, the flow is usually unsteady and non-uniform and, if it is separated into time average, periodic and random components, these may have different directions. Indeed, oscillatory components with different periods may have quite different directions. At the same time the sediment may contain a mixture of densities, sizes and shapes, which respond differently to the flow. For any given grain, its response to the flow depends on its position on an uneven bed. A grain at the crest of a ripple is much more exposed than one in the trough. For fine-grained (cohesive) sediments, the mobility may even depend on the history of consolidation and biological activity. All of these possible complications are to be kept in mind while the following text develops formulae for simplified scenarios.

6.3.2   Sediment characteristics

The mode and rate of sediment transport are determined by the parameters of the flow and of the sediment. Among the parameters of the sediment one thinks usually first of the size, which is an explicit part of most sediment transport formulae. The definition of "the size" is however not obvious for a mixture of grains with different shapes.

To this end, odd-shaped particles are usually assigned a diameter d, which is either the sieve diameter d_s, the volume diameter d_v or the fall diameter d_f, the latter two being the diameters of spheres with respectively the same volume and the same settling velocity. For relations between these different diameters see, e g, Fredsøe and Deigaard (1992) p. 197. The *dimensionless grain size* can be defined as $d\sqrt{(s-1)gd}\,/\,v$ which is a particle Reynolds number based on the nominal settling velocity $\sqrt{(s-1)gd}$. Mixtures of different diameters are primarily described in terms of statistics like the mean diameter $\bar{d}$ or the median diameter d_{50}.

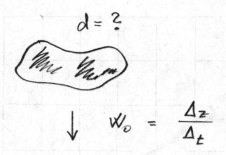

Figure 6.3.1:
The settling velocity is easier to define experimentally than the diameter for a natural sand grain.

The gradation, i e, the size variation is most simply quantified as the ratio between two size fractiles, e g, d_{90}/d_{10}. This ratio is large for well graded sediments and close to unity for well sorted sediments.

The sediment behaviour will also depend on its density ρ_s or the specific weight $s=\rho_s/\rho$. Typical beach sands consisting of quartz or carbonate have $\rho_s\approx2650\text{kg}/\text{m}^3$ corresponding to $s\approx2.65$.

Some sediment transport formulae include the angle of repose φ, which is also a friction coefficient between layers of sediment particles. It depends on the grain shape, gradation and packing of the sediment.

The solids concentration [solids volume/total volume] in the non-moving bed c_{max} depends on grain shape, gradation and packing. For spheres of uniform size the loosest packing is the cubic configuration for which the solid concentration is $c\approx0.52$, and the densest is the tetrahedron configuration for which $c\approx0.74$. For natural sands the range is $c_{max}\in[0.6;0.9]$.

The settling velocity w_o of a single particle in still water is easier defined than "the size". It does, however, depend on water temperature (via the fluid viscosity v) as well as on particle size and shape. For a spherical particle, the settling velocity is related to the diameter through

$$w_o = \sqrt{\frac{4(s-1)gd}{3C_D}}$$ (6.3.1)

where the drag coefficient C_D for spheres may be estimated from Figure 6.3.2.

An example of the settling velocity-diameter relationship for a natural beach sand sample, including odd-shaped particles, is shown in Figure 6.3.3.

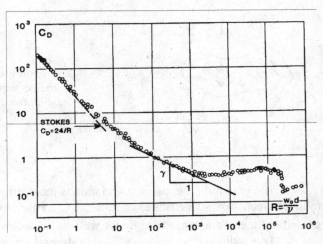

Figure 6.3.2:
Drag coefficients for spheres as functions of Reynolds number. The dashed line corresponds to Stokes Law, which gives $w_0 \sim d^2$ for laminar conditions.

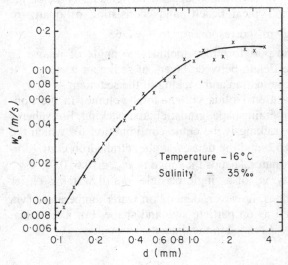

Figure 6.3.3:
Settling velocity versus sieve diameter for the various sieve fractions of one natural surf zone sand sample. The flattening of the curve for $d > 1$mm is due to the larger particles being disk shaped shell fragments. The behaviour of the smaller particles (<0.5mm) is fairly typical for quasi-spherical quartz sand.

218

She et al. (2005) showed that C_D for natural sand is typically two to three times larger than for spheres and suggested the formula

$$C_D = 2.1 + 40/R \qquad (6.3.2)$$

based on the sieve diameter.

Settling velocity depends on sediment concentration, the clear water value w_o being the limit for $c \to 0$. The other limit is the fluidization velocity w_f (see below), which can be seen as the settling velocity corresponding to the concentration c_{max} in the non-moving bed.

In the absence of fluid velocity shear it may be assumed that $w = w(c)$ varies as a simple power function between these two limits, i e,

$$\frac{w}{w_o} = (1-c)^n \qquad \text{for} \quad 0 < c < c_{max} \qquad (6.3.3)$$

This form has no theoretical basis but it is convenient and not seriously compromised by the available data from settling columns. Such experiments yield the typical exponent range $n \in [3; 4]$ for beach sand. Equation (6.3.3) does however not apply in boundary layer flow scenarios where the effects of velocity shear are dominant, as pointed out by Nielsen et al. (2002).

The porosity of a given sediment sample is the ratio

$$\varepsilon = \frac{\text{void volume}}{\text{solids volume}} \qquad (6.3.4)$$

The *permeability P* of a granular medium is usually an increasing function of the porosity but also a function of particle shapes and grading. It is usually determined experimentally via the *hydraulic conductivity K* which, for the x-direction is defined by

$$u = \frac{dq_x}{dA} = -K_x \frac{\partial}{\partial x}(\frac{p}{\rho g} + z) \qquad (6.3.5)$$

and similarly for the y- and z-directions. K_x, K_y, and K_z will be mutually different unless the porous medium is isotropic in which case the simple vector relation

$$\bar{u} = \frac{d\bar{q}}{dA} = -K \; \overline{grad(\frac{p}{\rho g} + z)} \tag{6.3.6}$$

called Darcy's Law is valid. The hydraulic conductivity is not a property of the porous medium alone since it depends on the density ρ and the kinematic viscosity ν. Based on a measured hydraulic conductivity, the permeability can be determined from

$$P_x = K_x \frac{\mu}{\rho g} \tag{6.3.7}$$

and similarly for the other directions, i e, the permeability has dimension of area while hydraulic conductivity has dimension of velocity.

As a rule of thumb, for uniform almost spherical particles, the hydraulic conductivity can be estimated roughly as

$$K \approx 10^{-3} \frac{gd^2}{\nu} \tag{6.3.8}$$

but note that K varies strongly with particle shape, gradation and packing.

The fluidization velocity w_f is the upward seepage velocity at which quicksand is formed, that is for which the upward pressure force equals the weight. Considering a unit volume on which the vertical pressure force is $-\frac{dp}{dz}$, and noting that according to Darcy's law for the vertical direction

$$\frac{w}{K_z} = -\frac{\partial}{\partial z}(\frac{p}{\rho g} + z),$$ where K_z is the "vertical" hydraulic conductivity,

this means

$$-\frac{dp}{dz} = \rho g(\frac{w_f}{K} + 1) = \rho g(sc_{max} + [1 - c_{max}]) \tag{6.3.9}$$

and hence

$$w_f = (s-1) c_{max} K \tag{6.3.10}$$

The erodability and settling behaviour of very fine sediments ($d < 50 \mu m$) is also dependent on *cohesion*, which originates from electrical attraction between the particles and depends on the chemical composition of the water as well as on the mineralogy and consolidation history of the sediment. See, e g, Winterwerp and van Kesteren (2004). A more detailed discussion of sediment parameters can be found in Allen (1984).

6.3.3 Forces on sediment particles

The forces which act to move the sediment are mainly drag forces $F_D \sim u^2 d^2$, lift forces $F_L \sim u^2 d^2$, and pressure forces $\overrightarrow{F_p} \sim -d^3 \overrightarrow{grad}\, p$.

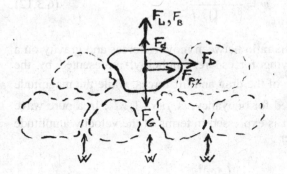

Figure 6.3.4: Forces on a sediment particle: Buoyancy force F_B, drag force F_D, gravity force F_G, lift force F_L, pressure force F_P, and seepage force F_S.

The definition of "The velocity" is not a trivial matter in the presence of the boundary layer and sheltering from other particles. We shall be using the free steam velocity $u_\infty(t)$ as a nominal drag velocity.

Similar issues arise with respect to the horizontal pressure gradient. Here the practical, and quite reasonable, approach is to assume hydrostatic conditions throughout the layer of moving sediment. The horizontal pressure gradient is then proportional to the free stream fluid acceleration, $\dfrac{dp}{dx} = -\rho \dfrac{du_\infty}{dt}$ and hence

$$F_{p,x} \propto -\rho \frac{du_\infty}{dt} d^3 \qquad (6.3.11)$$

Vertical pressure gradients give rise to the familiar buoyancy force $F_B = F_{p,z} \propto \rho g d^3$.

A vertical seepage velocity w corresponds to a vertical force, which may generate quicksand or alternatively stabilize the bed. Through Darcy's law ($\vec{u} = -K\,\overrightarrow{grad}\, p / \rho g$), the vertical seepage force per unit volume is $F_s = \rho g w / K$, where K is the hydraulic conductivity of the bed material.

Stabilizing forces are usually the gravity force $F_G \sim \rho s g d^3$ and the cohesive forces.

For a sloping sand surface, slope $\tan\beta$, gravity delivers a downhill tangential force of magnitude $F_G \sin\beta$ and the normal force due to gravity is reduced to $F_G \cos\beta$.

221

6.3.4   The mobility number
The mobility number

$$\psi = \frac{|u|^2}{(s-1)gd} \qquad (6.3.12)$$

is the simplest measure of the ratio between moving forces and gravity on a sediment particle. The moving forces are, implicitly, represented by the canonical form $F \propto \rho d^2 |u|^2$ of the drag and lift forces, while the magnitude of the gravity force, corrected for buoyancy, is $\rho(s-1)gd^3$. In a pure wave motion, the mobility number is expressed in terms of the velocity amplitude just above the boundary layer

$$\psi = \frac{(A\omega)^2}{(s-1)gd} \qquad (6.3.13)$$

Simplicity makes the mobility number very useful. However, the neglect of the different natures of the drag vs pressure forces is sometimes significant. For oscillatory flows, e g, $u(t) = A\omega \cos \omega t$, the relative importance of the pressure force, i e, F_p/F_D is measured by d/A, the inverse of the Keulegan-Carpenter number.

6.3.5   The Shields parameter
The Shields parameter

$$\theta(t) = \frac{\tau_b(t)d^2}{\rho(s-1)gd^3} = \frac{\tau_b(t)}{\rho(s-1)gd} = \frac{u_*^2(t)}{(s-1)gd} \qquad (6.3.14)$$

is a more sophisticated measure of sediment mobility which, in principle, accounts for both pressure gradients and drag, inasmuch as the acceleration effects, are incorporated in $\tau_b(t)$. However, as mentioned in Section 6.2 above, our experimental knowledge about $\tau_b(t)$ on movable sand beds, is restricted to rippled beds [Löfquist (1986)]. No measurements of $\tau_b(t)$ are, so far, available for flat, movable sand beds. Due to this shortage of information about the details of $\theta(t)$, other related quantities like the peak value

$$\hat{\theta} = \frac{\hat{\tau}}{\rho(s-1)\,gd} = \frac{\frac{1}{2}f_w(A\omega)^2}{(s-1)\,gd} = \frac{u_{*max}^2}{(s-1)gd} \qquad (6.3.15)$$

are often used for simplicity.

Like the shear stress (Equation (6.2.1)), the Shields parameter is often split into a form drag component and a skin friction component, $\theta = \theta'' + \theta'$ where only θ' is assumed to generate sediment transport. The questions about defining θ' and the hydraulic roughness for calculating it are usually bypassed by using a nominal grain roughness Shields parameter, based on $r = 2.5d_{50}$ instead. That is, instead of θ', the grain roughness Shields parameter $\theta_{2.5}$ with the peak (maximum) value

$$\hat{\theta}_{2.5} = \frac{\frac{1}{2}f_{2.5}(A\omega)^2}{(s-1)\,gd} \qquad (6.3.16)$$

is often used. The grain roughness friction factor $f_{2.5}$ is calculated from Equation (6.2.9).

The complex issue of hydraulic roughness of natural seabeds is comprehensively discussed in Nielsen (1992), Section 3.6.

6.3.6 Infiltration effects

In the swash zone, in particular, but possibly in the nearshore area in general (Conley and Inman 1992), the seepage forces associated with in- or outflow of water from the bed may influence the sediment mobility.

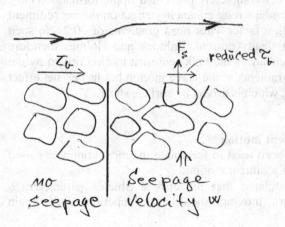

Figure 6.3.5: The effects of a positive seepage velocity are to Firstly: thicken the boundary layer and hence reduce the bed shear stress. Secondly to destabilize the particles through the upward seepage force F_s. For infiltration ($w<0$) both effects are reversed.

223

Flow perpendicular to the bed has two opposing effects on sediment mobility, see Figure 6.3.5. That is, inflow will make the boundary layer thinner and thereby increase the shear stress. At the same time however, the inflow exerts a downward drag, which helps to stabilize the sediment.

Martin (1970) found experimentally that either of the two effects may be dominant, depending on the sediment size and density. For large or very dense particles, the mobilizing effect of the increased shear stress dominates. For finer and/or lighter sediments the stabilizing effect of the downward drag dominates.

Nielsen et al. (2001) suggested incorporating the influence of a vertical seepage velocity w (positive upwards) on sediment stability through a modified Shields parameter

$$\theta = \frac{u_{*_o}^2(1-\alpha\frac{w}{u_{*_o}})}{\left[s-1-\beta\frac{w}{K}\right]gd} \qquad (6.3.17)$$

where u_{*_o} is the shear velocity with no seepage, α and β are dimensionless coefficients and K is the hydraulic conductivity of the sand. Based on experiments with non-breaking waves over a horizontal bed, Nielsen et al. (2001) recommended $(\alpha,\beta) = (16, 0.4)$.

The infiltration effects in the experiments of Nielsen et al. (2001) $[(h,H) = (0.45\text{m}, 0.15\text{m})]$ were relatively weak. However, Baldock and Holmes (1998) performed smaller scale $[(h,H) = (0.1\text{m}, 0.038\text{m})]$ experiments, with similar (0.2mm quartz) sand and found that the stabilizing infiltration effect could dominate over these waves. They found that vertical head gradients greater than 0.1 completely prevented ripple formation. They also found that exfiltration had a strong enhancing effect on the net sediment transport rate by waves, e g, a factor 4 for head gradients of –0.2. No such increase was found with a steady current. Baldock and Holmes therefore concluded that exfiltration enhances the bulk sediment motion driven by the strong horizontal pressure gradients in the wave motion but has no net effect on the transport by currents, which is driven by surface shear.

6.3.7   Initiation of sediment motion

The Shields parameter has been used to formulate the most commonly used criterion for the initiation of sediment motion.

Shields (1936) postulated that the critical Shields parameter θ_c at which the sediment starts moving, should be a function of the grain

Reynolds number u_*d/v. This is reasonably well supported by data as indicated by Figure 6.3.6, as long as the specific gravity s does not vary too much.

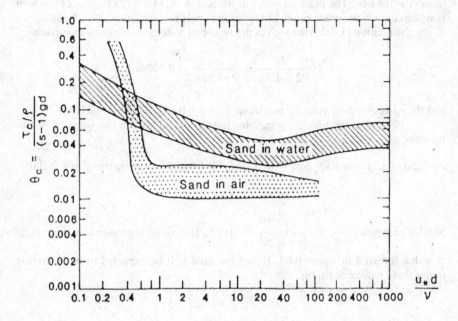

Figure 6.3.6: Within a given, reasonably narrow, range of specific gravity, the critical Shields parameter θ_c is a function of the grain Reynolds number only. For sand in water the 'rule of thumb' value is 0.05.

However, if vastly different s-values are considered, e g, 2.65 for quartz in water and 2200 for quartz in air the dependence of θ_c upon s is quite strong. For coastal applications, recent work by Terrile et al. (2006) indicates that wave shape has strong and as yet not fully understood influence on the initiation of sediment motion.

Cohesive sediments are much more resilient to erosion than indicated by the Shields diagram. However, the presence of sand may enhance the erosion process very significantly as described by Mitchener and Torfs (1996).

225

Example 6.3.1:
Find out if 0.3mm sand is likely to be moved by 9s waves with height 2m in 25m of water.

First we need to determine the parameters of the wave induced water motion at the bed. The dimensionless depth is $k_o h = (4\pi^2/gT^2) h = 1.243$ which is within the 1% accuracy range of Equation (1.4.16) which gives $kh = 1.388$.

Equation (1.4.28) then gives the near-bed water particle semi-excursion

$$A = \frac{H}{2\sinh kh} = \frac{2m}{2 \times 1.880} = 0.532m$$

and the corresponding velocity amplitude is $A\omega = 0.532 \times (2\pi/9) = 0.371 m/s$.

The relevant Shields parameter is then calculated on the basis of "grain roughness":

$$r = 2.5d, \text{ i e, } f_{2.5} = \exp\left[5.5\left(\frac{2.5d}{A}\right)^{0.2} - 6.3\right] = 0.0081. \text{ Subsequently, we find the}$$

Shields parameter $\hat{\theta}_{2.5} = \dfrac{\dfrac{1}{2} f_{2.5} (A\omega)^2}{[s-1] gd} = 0.12$. This value is greater than the typical

θ_c value for sand in water: 0.05. Hence the sand will be expected to be in motion under these wave conditions.

☺

6.4   BEDLOAD AND SHEET FLOW

6.4.1   Introduction

Sediment transport over flat beds under non-breaking waves is concentrated in a very thin layer (of the order 100 grain diameters) around the level of the undistributed bed, see Figure 6.4.1. The process is called sheet flow and may consist partly of bedload and partly of suspended load. Measurements from within this layer were first provided by the pioneering work of Horikawa et al. (1982) but more details have been added by the work of Dick and Sleath (1991a,b), Zala-Flores and Sleath (1997), Sleath (1999) and Ribberink et al. (1994, 2000), O'Donoghue and Wright (2004) and Watanabe and Sato (2004).

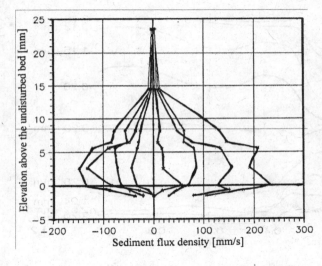

Figure 6.4.1: Instantaneous sediment flux densities measured in a water tunnel by Ribberink et al. (1994). The grain diameter is 0.21mm, the wave period 7.2s and the peak velocities 1.8m/s and 1.4m/s in the landward and seaward directions respectively.

Particularly interesting with respect to future model developments are the time series of concentrations at and below $z=0$, see Figure 6.4.2. These ought to help bring the state of the art beyond the current use of separate bedload models and suspension models with nominal reference concentrations and into the era of continuous descriptions covering all moving sediment from the lowest level of sediment motion to the water surface. A number of detailed models of the granular flow near $z=0$ have been developed with this aim, e g, Hanes and Bowen (1985) and Kaczmarek and Ostrowski (2002). However, a great many simplifying and sometimes unrealistic assumptions are still required in order to make them workable.

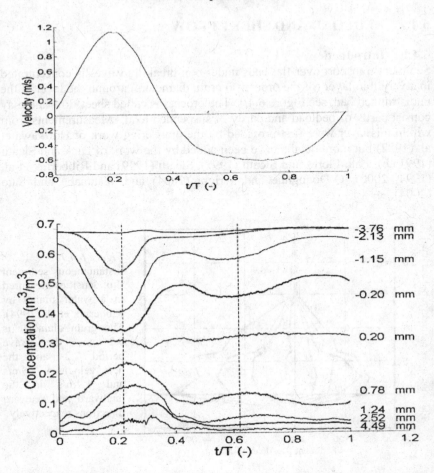

Figure 6.4.2: Free stream velocity and sediment concentrations in sheet flow under waves in a flume, measured by Ribberink et al. (2000). The wave period is 6.5s and $d_{50} = 0.245$mm. At levels below the undisturbed bed, the peak velocities create a dip in concentrations. Above the undisturbed bed level they create peaks. No sediment movement occurs at $z = -3.76$mm and below.

6.4.2   What is bedload?

In the following, the term bedload refers to that part of the sediment load, which is carried by inter-granular forces, as opposed to suspended load and wash load, which are carried by fluid drag. This definition which is due to Bagnold (1956) is not convenient from a measurements point of view since a given sand particle may be supported by a combination of the two kinds of forces and hence be partly bedload, partly suspended load. It is however the most convenient definition with respect to conceptual model building.

6.4.3 Dispersive stress and the amount of bedload

Consider a sand bed, which is exposed to a bed shear stress τ_b capable of mobilizing the top layer of sand. Given that the shear stress on the next layer is then at least as great as τ_b, one might expect that it will also be mobilized, and so will the next, and so on. This is, however, in conflict with observations: Usually only a finite number of layers will be mobilized.

This represented a paradox until Bagnold (1956) introduced the concept of dispersive stress. That is, the weight of the bedload is transferred as an effective normal stress σ_e to the non-moving bed, and this extra effective stress, enables the topmost non-moving layer to withstand a shear stress greater than τ_c.

Consider the situation in Figure 6.4.3 and assume that the maximum shear stress, which can be withstood by the topmost non-moving layer of particles is given by the Coulomb criterion

$$\tau_{max} = \tau_c + \sigma_e \tan\phi = \tau_c + \tan\phi \int_{z_b}^{\infty} \rho g(s-1) c_b dz \qquad (6.4.1)$$

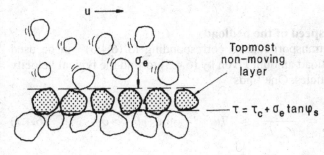

Figure 6.4.3: Due to the weight of the bedload above, which is transferred as the dispersive stress σ_e (Bagnold 1956), the topmost non-moving layer can withstand a shears tress that is greater than τ_c.

Then, quantifying the total amount of bedload (m^3/m^2) by its equivalent resting thickness at concentration c_{max}

$$L_b = \frac{1}{c_{max}} \int_{z_b}^{\infty} c_b dz \qquad (6.4.2)$$

we find

$$L_b = \frac{\tau - \tau_c}{\rho g (s-1) c_{max} \tan \phi} \tag{6.4.3}$$

or, introducing the Shields parameter and using Bagnold's (1956) general ballpark value

$$c_{max} \tan \varphi \approx 0.4 \tag{6.4.4}$$

we get

$$L_b \approx 2.5 (\theta' - \theta_c) d \tag{6.4.5}$$

6.4.4 Meyer-Peter type bedload formulae

Meyer-Peter and Müller (1948) found that steady bedload was well described by

$$\Phi_b = \frac{q_b}{d\sqrt{(s-1) \, gd}} = 8 \, (\theta' - \theta_c)^{1.5} \tag{6.4.6}$$

and a number of similar formulae have since been derived from physical considerations, cf Madsen (1991) and Fredsøe (1993). Correspondingly, the net transport from half a sine wave $\Phi_{T/2}$ was found by Nielsen (1992) to behave similarly, see Figure 6.4.4.

6.4.5 The typical speed of the bedload

The semi-empirical transport rate q_b corresponding to (6.4.6) can be used together with the bedload amount given by (6.4.5) to find the typical velocity $<u_b>$ for bedload particles. One finds

$$\langle u_b \rangle = \frac{q_b}{c_{max} L_b} \approx 7 u_* \qquad \text{for} \quad \theta' >> \theta_c \tag{6.4.7}$$

6.4.6 Energy dissipation and bedload

In order to bypass many of the difficult details of the sediment transport process, Bagnold (1963) suggested the shortcut of relating the sediment transport rate q_s [m²/s] directly to the energy dissipation rate D_E [W/m²]. The basic idea was that the power P_b per unit bed area expended on dragging the bedload along should be a certain fixed fraction ε of D_E. The required drag power P_b is estimated as

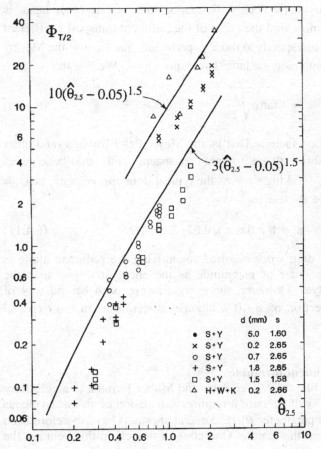

$$10(\hat{\theta}_{2.5} - 0.05)^{1.5}$$

$$3(\hat{\theta}_{2.5} - 0.05)^{1.5}$$

	d (mm)	s
• S+Y	5.0	1.60
× S+Y	0.2	2.65
○ S+Y	0.7	2.65
+ S+Y	1.8	2.85
□ S+Y	1.5	1.58
△ H+W+K	0.2	2.66

Figure 6.4.4:
The total net transport from half a sine wave can be described in terms of $\hat{\theta}_{2.5}$ and a Meyer-Peter and Müller type of transport formula. The coefficient will however depend on the amount of suspended load, which is greater for fine sediments at a given $\hat{\theta}_{2.5}$.

$$P_b \approx \rho \, (s-1) \; g \; c_{max} L_b \tan \varphi \; \langle u_b \rangle = \rho \, (s-1) \; g \tan \varphi \; q_b \qquad (6.4.8)$$

where the first part is recognized as the frictional force due to the submerged weight of bedload per unit area while $\langle u_b \rangle$ is, somewhat arbitrarily but conveniently, taken as the relevant dragging velocity. Inserting the Meyer-Peter expression for q_b then gives

$$P_b \approx \rho (s-1) \; g \tan \phi \sqrt{(s-1)gd} \; d \; 8(\theta' - \theta_c)^{1.5} \qquad (6.4.9)$$

The boundary layer energy dissipation rate, on the other hand, is $D_E = \overline{\tau_b u_\infty}$. We can now find the value of the sediment transport coefficient $\varepsilon = P_b / D_E$, which corresponds to those experiments that support the Meyer-Peter formula, by combining the latter two expressions. We find that

$$\varepsilon \approx 8 \tan\phi \sqrt{\frac{f}{2}} \qquad \text{for } \theta' >> \theta_c \qquad (6.4.10)$$

where f is the friction factor. That is, the Meyer-Peter formula (and other formulae of the form $\Phi \propto \theta^{1.5}$) are consistent with the "energetics hypothesis" $P_b = \varepsilon D_E$. Taking $<u_b>$ as the typical dragging velocity, $\tan\phi \approx 0.6$ and $f \approx 0.02$ we are lead to

$$\varepsilon \approx 8 \times 0.6 \times \sqrt{0.02/2} \approx 1/2 \qquad (6.4.11)$$

That is, the drag work required for moving the sediment along is usually of the same order of magnitude as the energy dissipation in the bottom boundary layer. However, the energetics approach has no way of determining the direction of $\overline{q_s}$ if u changes direction as in most coastal flows.

6.4.7 Velocity moment formulae

Adapting a formula like the Meyer-Peter and Müller formula to an arbitrary unsteady flow is difficult because it requires calculation of an instantaneous skin friction Shields parameter $\theta'(t)$. Several shortcuts have therefore been tried in order to avoid this hurdle. One class of models in this vein are the velocity moment formulae where an empirical relation of the form $\overline{q_s} = K \overline{|u_\infty|^{p-1} u_\infty}$ is sought, where u_∞ is the velocity just above the bottom boundary layer, and the overbar denotes time averaging. Ribberink and Al-Salem (1994) showed that for 0.21mm sand and typical prototype wave-current (collinear) flows, such a model with $p = 3$ and $K = 0.00018 s^2/m$ gave good predictions. However, similar experiments by Ribberink and Chen (1993) using 0.13mm sand could not be modelled in this way. For the finer sand, the assumption of quasi steadiness $q_s(t) \propto |u_\infty(t)|^{p-1} u_\infty(t)$, which seems valid for 0.21mm (and probably coarser) sand, breaks down as it does for most sand sizes over rippled beds. The reason for the breakdown of the quasi-steady approach is the presence of a considerable amount of suspended sand and that suspended sand does not move in step with $u_\infty(t)$. The Ribberink and Al-Salem sheet flow formula

$$\overline{q_s} = 0.00018 \ \overline{u_\infty^3} \ \text{(SI units)} \tag{6.4.12}$$

can be tied in with the Meyer-Peter and Müller type formulae by considering a particular velocity variation, e g, half a sine wave, and sand of quartz density: $s = 2.65$. In this case it corresponds to

$$\Phi = 0.003\overline{\psi^{3/2}} \tag{6.4.13}$$

and, if a constant friction factor of $f_w \equiv 0.008$ is assumed, it is equivalent to the Meyer-Peter and Müller like formula $\Phi = 12\overline{\theta^{3/2}}$ which was shown to fit the 0.2mm sand data in Figure 6.4.4.

6.4.8 Bedload and sheetflow transport in an arbitrary flow

For quartz sand which is not too fine ($d_{50} \geq 0.2$mm) the Equation (6.4.12) has been shown to apply for water tunnel experiments with combinations of currents and sine waves and for currents plus waves with Stokes asymmetry. However, as pointed out by Ribberink et al. (2000), it under-predicts the sediment transport rates measured under real waves by about a factor 2, see Figure 6.4.5.

We see that the flume waves generate about twice the sediment transport rate for a given value of $\overline{u_\infty^3}$ compared with the "sine waves plus current" or the 2nd order Stokes waves in the tunnel. There are two possible reasons for this difference:

Firstly, the flume waves generate a steady drift at the bed and a corresponding mean bed shear stress in the shoreward direction, Figure 1.4.10. This streaming is driven by Reynolds stress terms of the form $-\rho\overline{\tilde{u}\tilde{w}}$, which are non-zero in real waves with a bottom boundary layer as explained by Longuet-Higgins (1956). These stresses do not exist in the water tunnels where $\tilde{w} \equiv 0$.

For the flume situation, the streaming related extra bed shear stress can be estimated from

$$-\rho(\overline{\tilde{u}\tilde{w}})_\infty = \rho\frac{1}{4\sqrt{2}}kA^3\omega^2 f_w \tag{6.4.14}$$

with f_w based on the hydraulic roughness expression (6.5.17), and the corresponding contribution to the Shields parameter is

$$\Delta_\theta = \frac{-(\widetilde{u}\widetilde{w})_\infty}{(s-1)gd} \tag{6.4.15}$$

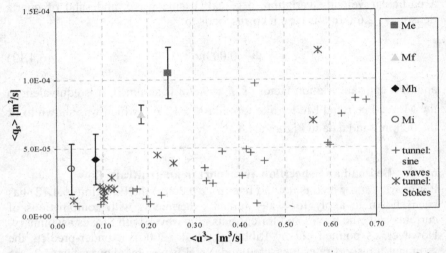

Figure 6.4.5: Measured net sediment transport rates versus third moment of the free stream velocity for real waves in a flume (symbols with error bars) and from water tunnel experiments with "sine waves plus currents", +, and 2nd order Stokes waves, ∗.

Secondly, the flume waves have a bit of saw-tooth asymmetry with more abrupt acceleration before the wave crest. The effect of this saw-tooth asymmetry on the bed shear stress was discussed in connection with Figure 6.2.6 and accounted for by Equation (6.2.10). The corresponding instantaneous grain roughness Shields parameter is

$$\theta_{2.5}(t) = \frac{\frac{1}{2}f_{2.5}}{(s-1)gd_{50}}\left[\cos\phi_\tau u_\infty(t) + \sin\phi_\tau \frac{u_\infty(t+\delta_t) - u_\infty(t-\delta_t)}{2\omega_p\delta_t}\right]^2 \text{sign}\left(u_*(t)\right) \tag{6.4.16}$$

where the instantaneous friction velocity is taken as

$$u_*(t) = \sqrt{\frac{1}{2} f_{2.5}} \left[\cos\phi_\tau u_\infty(t) + \sin\phi_\tau \frac{u_\infty(t+\delta_t) - u_\infty(t-\delta_t)}{2\omega_p \delta_t} \right] \quad (6.4.17)$$

When the total grain roughness Shields parameter $\theta_{2.5}(t)$, obtained as the sum of (6.4.15) and (6.4.16), is inserted into the Meyer-Peter and Muller style transport formula we get

$$\Phi(t) = \frac{q_s(t)}{\sqrt{(s-1)gd_{50}^3}}$$

$$= \begin{cases} 0 & \text{for} \quad \theta_{2.5} < 0.05 \\ 12 \ [\theta_{2.5}(t) - 0.05]\sqrt{\theta_{2.5}(t)} \ \text{sign}(u_*(t)) & \text{for} \quad \theta_{2.5} > 0.05 \end{cases} \quad (6.4.18)$$

which is in good agreement with the flume measurements of Ribberink et al. (2000), see Figure 6.4.6. For further details and comparisons with data see Nielsen (2006).

The method outlined above also gives reasonable agreement with measured swash zone net sediment transport rates, cf Nielsen (2002).

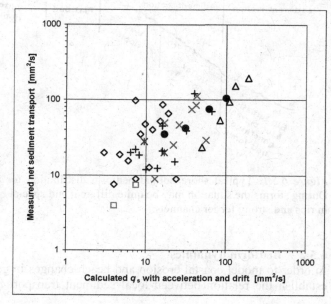

Figure 6.4.6: Measured versus calculated net sediment transport rates for the same data as in Figure 6.4.5.

6.5   BEDFORMS AND HYDRAULIC ROUGHNESS

6.5.1   Introduction
Bedforms have a strong effect of sediment transport through modifications to the flow and to the mechanisms of sediment entrainment. The nature of the influence is different for different types of bedforms. Thus, bars will determine the position of the breakpoint and the longshore location of rip currents while vortex ripples influence the vertical structure of the flow and of sediment concentrations. A schematic overview of typical coastal bedforms on one type of beach morphology is provided by Figure 6.5.1. For an overall description of coastal bedforms see also Clifton (1976).

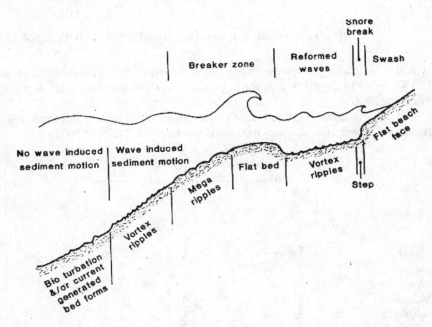

Figure 6.5.1: Typical shore normal bedform distribution for an accreting beach. During storms the situation may be quite different and special bedforms are present in rips and strong feeder channels.

6.5.2   Bedform dynamics
In order to model coastal erosion and beach changes in general, we need to establish the relation between local sediment transport rates and bedlevel

changes. This is done in the following section. It turns out that the obvious requirement of conservation of sediment volume leads to a simple relationship between erosion rates and the spatial gradients of sediment transport. The same simple continuity equation is also used to gain qualitative insights into the conditions for bedform growth and migration.

6.5.2.1 Observations of migrating natural bedforms

From analysis of beach profiles from Duck N C, Larson and Kraus (1992) found offshore bar migration speeds up to 18m/day and onshore bar speeds up to 2.9m/day. They noted that the offshore speeds in particular were low estimates because surveys were usually not done immediately after high energy events. In other words, their speeds were averages between surveys rather than peak values during storms. While Larson and Kraus reported no long-term trend, Wijnberg (1996) described how, on the multibarred Dutch coast, all bars are moving offshore in the long term with speeds of the order 50m/year. Elgar et al. (2001) demonstrated that onshore bar migration during moderate wave conditions can be understood in terms of wave acceleration skewness. When storm waves arrive however, the breakpoint moves seaward and strong undertow makes the bar move seaward towards the new breakpoint. Mearipples have been observed to move landward with typical speeds of 0.5cm/min ranging up to about 2.5cm/min by Gallager et al. (1998) and Gallagher (2003). Inman and Bowen (1963) observed both seaward and landward ripple migration at up to 2mm/min in the laboratory, and similar speeds, i e, typically 0.5 to 1mm/min, were observed by Farachi and Foti (2002) and Davis et al. (2004). In the field, Dingler (1974) observed ripples migrating landwards at rates up to a few cm/min, while Traikovsky et al. (1999) found a peak value of 80cm/day with an average of 24cm/day over a 25day period. This ripple migration is due to a combination of wave asymmetry, $\overline{u^3} \neq 0$, acceleration asymmetry, $\overline{(\frac{du}{dt})^3} \neq 0$, and boundary layer streaming. All of these are greater in shallower depths for given (H,T) which explains the much greater migration rates observed by Dingler at $h \approx 2$m compared with Traikovsky et al. at $h \approx 11$m.

6.5.2.2 The continuity equation

The relation between changes in the sediment transport rate and changes in bed elevation is derived by expressing the conservation of sediment volume for a control volume like the one in Figure 6.5.2.

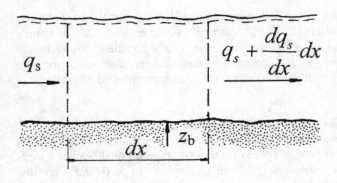

$$q_s + \frac{dq_s}{dx}dx$$

q_s

z_b

dx

Figure 6.5.2:
2D control volume. The solids fraction of the porous bed is n.

If there is more sediment leaving than entering, the bed level of the control volume must go down and if the solids fraction of the bed is n, the simple 2D continuity equation is

$$n\frac{dz_b}{dt} = -\frac{dq_s}{dx} \qquad (6.5.1)$$

while the more general 3D version, where the sediment transport rate is a vector in the xy-plane, $\vec{q_s} = (q_{s,x}, q_{s,y})$ reads

$$n\frac{dz_b}{dt} = -div\,\vec{q_s} = -\frac{dq_{s,x}}{dx} - \frac{dq_{s,y}}{dy} \qquad (6.5.2)$$

i e, the local erosion rate is $\frac{1}{n}div\,\vec{q_s}$. Both of these equations are approximate in that they ignore changes to the amount of sediment suspended above z_b inside the control volume.

6.5.2.3 Bedforms migrating with constant form

The continuity principle can be used very easily to derive sediment transport rates from the shape and speed of (2D) bedforms if these are assumed to migrate with constant form. The argument is as follows. Consider two dimensional bedforms of arbitrary shape f, which propagate with speed c so that the sand level can be described by

$$z_b(x,t) = f(x-ct) \tag{6.5.3}$$

Inserting this into the lefthand side of (6.5.1) we get

$$n(-c)f'(x-ct) = -\frac{dq_s}{dx} \tag{6.5.4}$$

and integration with respect to x gives

$$q_s = q_o + ncf(x-ct) = q_o + ncz_b(x,t) \tag{6.5.5}$$

where the constant of integration q_o is the sediment transport rate at the point(s) where the bed level is zero.

Equation (6.5.5) shows that if the bedforms move downstream ($c>0$) the sediment transport rate varies in step with z_b, i e, the sediment transport rate is maximum over the bedform crest and minimum over the trough. Conversely, for bedforms which move upstream like antidunes, and thus have $c<0$, the maximum sediment transport rate occurs over the trough and the minimum over the bedform crest.

Experiments show that both ripples and megaripples migrate with typical speeds around $c \approx 1$cm/min in surf zones with skewed velocities. This fairly constant value of the migration speed c indicates that there is a fairly constant ratio between bedform height and bedform related sediment transport for these types of bedforms. That is, Equation (6.5.5) gives

$$q_{s,\text{crest}} - q_{s,\text{trough}} = nc\eta \approx \frac{2}{3}(1\,\text{cm}/\text{min})\eta \tag{6.5.6}$$

for ripples and megaripples under skewed nearshore waves.

6.5.2.4 Migration and growth of sinusoidal bedforms

Making use of the simple continuity principle, we saw above that if bedforms migrate with constant form, sediment transport rates and bed shape are exactly in step for $c>0$ and exactly opposite if $c<0$. The same simple continuity principle can be used to show that if q_s is out of step with z_b as in Figure 6.5.3, the bedforms must be either growing or eroding. We shall see that, if q_s peaks before the crest, i e, $\delta_x<0$, the bedform will steepen. Conversely, if $\delta_x>0$ the crest will be eroding and the bedform becomes flatter.

239

The qualitative principle is quite general but the mathematics are only simple for a simple harmonic bedform, so consider

$$z_b(x,t) = A\cos[\frac{2\pi}{\lambda}(x-ct)] = A\cos k(x-ct) \qquad (6.5.7)$$

or by introducing the complex exponential $e^{ix} = \cos x + i\sin x$ and attaching physical meaning to the real part only

$$z_b(x,t) = A e^{ik(x-ct)} \qquad (6.5.8)$$

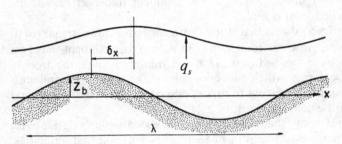

Figure 6.5.3: If the peak in q_s occurs after the crest ($\delta_x > 0$), the bedform will become flatter. Conversely if $\delta_x < 0$.

Assuming then that the sediment transport varies in a similar way to the bed elevation, as indicated in Figure 6.5.3

$$q_s(x,t) = q_o + q_1 e^{ik(x-ct-\delta_x)} \qquad (6.5.9)$$

the steepening or flattening rate of the bedform is determined by the continuity equation (6.5.1). Inserting the expressions for z_b and q_s we get:

$$nAik(-c)e^{ik(x-ct)} = -q_1 ik e^{ik(x-ct-\delta_x)} \qquad (6.5.10)$$

which we can solve for the speed of propagation and get

$$c = \frac{q_1}{nA}e^{-ik\delta_x} = \frac{q_1}{nA}[\cos k\delta_x - i\sin k\delta_x] \qquad (6.5.11)$$

240

This result is most easily interpreted when c is real-valued, i e, if $\delta_x=0$, leading to $c=\dfrac{q_1}{nA}$ and if $\delta_x=\lambda/2$ leading to $c=-\dfrac{q_1}{nA}$. That is, the bedforms move downstream with speed $c=\dfrac{q_1}{nA}$ for $\delta_x=0$ and upstream with speed $c=-\dfrac{q_1}{nA}$ for $\delta_x=\lambda/2$.

We can however also interpret the complex c-values which result for $\delta_x\neq 0,\lambda/2$. Then, the imaginary part of c expresses the growth rate. This is seen by inserting the solution $c = \dfrac{q_1}{nA}[\cos k\delta_x -i\sin k\delta_x]$ into $z_b(x,t) = Ae^{ik(x-ct)}$ which leads to

$$z_b(x,t) = Ae^{-k\frac{q_1\sin k\delta_x}{nA}t}\; e^{ik(x-\frac{q_1\cos k\delta_x}{nA}t)} \tag{6.5.12}$$

or

$$z_b(x,t) = Ae^{-k\,\text{Im}\{c\}t}\, e^{ik(x-\text{Re}\{c\}t)} = Ae^{-k\,\text{Im}\{c\}t}\cos k(x-\text{Re}\{c\}t) \tag{6.5.13}$$

where Re{} and Im{} denote real and imaginary parts of a complex number respectively.

The analysis above is of a unidirectional flow scenario. However, the derived relative growth rate

$$\frac{1}{A}\frac{dA}{dt} = -k\frac{q_1\sin k\delta_x}{nA} = -\frac{2\pi}{\lambda}\frac{q_1\sin\dfrac{2\pi\delta_x}{\lambda}}{nA} \tag{6.5.14}$$

can be used to see what it takes for symmetrical bedforms in a symmetrical oscillatory flow to grow. The general indication is that the bedforms will grow if $\sin\dfrac{2\pi\delta_x}{\lambda}<0$ corresponding to $-\lambda/2<\delta_x<0$. That is, if the sediment transport rate peaks along the uphill slope, i e , where the flow is contracting. This may be the case for both half-cycles of a symmetrical wave motion and the growth rate will then be proportional to the gross sediment transport rate $q_{s,gross}= |q_{s,landward}| + |q_{s,seaward}|$, not to the net transport rate. A corollary to this is that the direction of bedforms tends to be such that the gross transport rate across the crest is maximized. This is an experimentally verified fact from

241

Aeolian (Rubin and Hunter 1987) as well as from coastal flows, (Gallagher et al. 1998).

6.5.2.5 Numerical issues in bed updating

The nature of the continuity equation (6.5.1) is the source of much trouble for numerical modellers. This is related to the fact that, it becomes a so-called hyperbolic equation, if the sediment transport rate is proportional to the bed elevation, $q_s \propto z_b$ which, as we saw above, corresponds to bedforms which propagate with constant form.

Hyperbolic equations are often solvable analytically, but they are notoriously difficult to solve numerically. Simple numerical schemes give rise to spurious oscillations and smoothing with the aim of getting rid of these will often smooth out most of the real detail as well.

The "centered in space and forward in time" numerical scheme, which is often applied in coastal morphodynamics modelling, is a dispersive scheme and when applied to non-linear systems it becomes unconditionally unstable. To overcome this, higher order Lax-Wendroff schemes and/or smoothing have been applied with limited success. For a recent review see, e g, Callaghan et al. (2006).

Figure 6.5.4: Small vortex ripples in fine sand exposed at low tide. The profile of a vortex ripple in somewhat coarser sand in a wave flume is shown in Figure 6.5.6.

6.5.3 Vortex ripples

Vortex ripples are symmetrical bedforms which are present in wave dominated flows. Under low energy conditions, $\theta_{2.5} < 0.5$, the ripples have sharp crests which cause regular vortex shedding, Figure 6.5.6. Their length λ is then of the order $1.3A$ and their height η about $0.2A$. As the flow gets more vigorous the ripples get relatively shorter and flatter, they vanish at around $\theta_{2.5} = 1$.

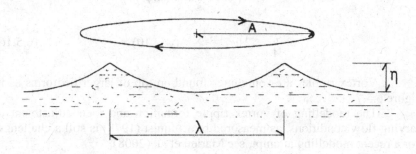

Figure 6.5.5: The length λ of wave ripples is related to the amplitude A of the water motion. Such a flow scale is not available for current ripples.

Figure 6.5.6: Sharp crested vortex ripples over which sand is entrained into the lee vortices, which form periodically at the crests.

The vortex ripples are larger (in terms of the relative measures λ/A and η/A) and more regular under regular waves than under random waves. Asymmetrical wave motion generates asymmetrical ripples with a milder slope facing the largest peak velocity. Quantitatively, the behaviour for field conditions (irregular waves) is described by Equations (6.5.15) and (6.5.16)

$$\frac{\lambda}{A} = \exp\left(\frac{693 - 0.37\left(\ln\psi\right)^{8}}{1000 + 0.75\left(\ln\psi\right)^{7}}\right) \qquad (6.5.15)$$

$$\frac{\eta}{A} = 21\psi^{-1.85} \qquad (\psi > 10) \qquad (6.5.16)$$

Vortex ripples are sometimes found on top of larger bedforms as in Figure 6.5.7.

The modelling of vortex ripple formation and their adjustment to varying flow conditions, as measured by Löfquist (1978) is still a challenge. For a recent modelling attempt, see Marieu et al. (2008).

Figure 6.5.7: Bedforms, very similar to river dunes in a rip feeder channel. At the top of the feeder, where the current is weak, the bedforms are shore-parallel vortex ripples. Further down where the current becomes dominant, one finds these dune-like bedforms, often with ripples on top.

6.5.4   Megaripples

Megaripples are larger than vortex ripples, typically $\lambda > 2A$, and have rounded crest which do not usually cause flow separation and vortex shedding. The flow boundary layer over megaripples is therefore for the most part similar to that over a flat bed. However, large (up to 1m high) clouds of suspended sand are often observed over particularly active spots where the three dimensional flow structure generates strong vortices.

6.5.5   Bedforms in combined wave current flows

Coastal flows are usually combinations of waves and currents with a variety of relative orientations. These different flow combinations result in a great variety of bedforms on which there is a substantial literature.

Within certain limits, the bedforms can be so dominated as to be wholly determined by either the waves or the current, see e g, Sleath (1984) p. 169 or van Rijn (1993) p. 5.45. Details for steady, intermediate conditions and for transitional states can be found in several experimental studies, e g, Amos et al. (1988), Arnott and Southard (1990) and Hay and Wilson (1994).

6.5.6   Hydraulic roughness

Natural sand beds are never perfectly flat and hence the Nikuradse roughness $r = d_{50}$ is usually an underestimate. For a nominally flat sand bed with little sediment movement we use $r = 2.5d_{50}$.

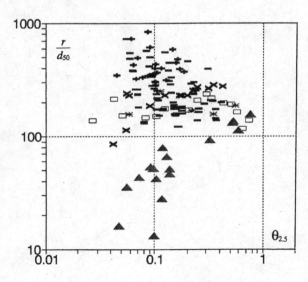

Figure 6.5.8:
Observed hydraulic roughness for oscillatory flows over flat beds = triangles, and rippled beds = all other symbols. Data from energy dissipation measurements by Carstens et al. (1968).

When the bed is rippled and/or when a considerable amount of sediment is in motion, and thus contribute to the momentum transfer from the fluid to the bed, the roughness is much greater as indicated by Figure 6.5.8. Those "measured" r-values are inferred from measurements of energy dissipation D_E from which a wave friction factor f_w can be calculated, see Nielsen (1992) p. 27. The roughness r is then found by using Equation (6.2.8) in reverse.

We note that under flow conditions which correspond to important sediment transport scenarios, the hydraulic roughness is usually of the order $100d_{50}$.

Based on the energy dissipation data set in Figure 6.5.8 and the corresponding bed-form data, Nielsen (1992) suggested the formula

$$r = 8\frac{\eta^2}{\lambda} + 170\sqrt{\hat{\theta}_{2.5} - 0.05}\, d_{50} \qquad (6.5.17)$$

6.6 THE MOTION OF SUSPENDED PARTICLES

6.6.1 Introduction

In order to model suspended sediment transport which occurs under breaking waves, over steep bedforms and in the swash, it is necessary to understand the behaviour of suspended particles in various types of flows. In particular, we need to know

(1) how fast do the particles settle and
(2) how closely do they follow the horizontal velocity of the water.

It turns out that the former question is very complex and has attracted several erroneous theories along with a few useful ones. We shall see how the dominant effects have little to do with the micro-dynamics of the flow (non-linear drag, Magnus effect etc.) but all to do with the coherent structures of the flow. The latter question has a simple answer: For all practical purposes, sediment particles have the same horizontal velocity as the surrounding fluid. Most of the results in the following hold analogously for buoyant particles and bubbles as for sinking sediments.

The overwhelming importance of the overall flow structure and the, sometimes surprising, nature of its influence is illustrated by Figure 6.6.1.

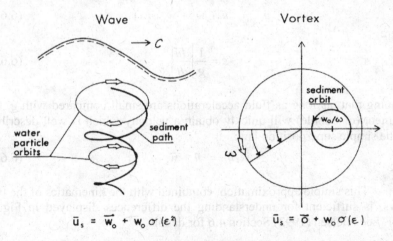

Figure 6.6.1: Wave motions and vortex motions are similar in the sense that both have elliptical water particle orbits. Yet, sediment particles behave very differently in the two. The vortex will trap particles (and bubbles) on closed orbits, while the wave motion has no such trapping ability. Settling is unhindered by waves (without turbulence).

6.6.2 Equation of motion for a suspended particle or bubble

The following discussion will neglect effects of particle-particle interactions, and of particle-boundary interactions but these simplifications are deemed to be of little consequence for sediment moving more than about one centimetre above the bed. We shall also neglect the viscous Basset history term and effects of particle spin. Experience shows that these effects contribute little to the most important phenomena. With these simplifications, the velocity vector $\vec{u}_s$ of a spherical sediment particle with diameter d and density ρ_s in a flow with velocity $\vec{u}$ is given by

$$\rho s \frac{\pi}{6} d^3 \frac{d\vec{u}_s}{dt} =$$

$$\rho s \frac{\pi}{6} d^3 \vec{g} - \frac{\pi}{6} d^3 \overline{\text{grad } p} + \frac{1}{2} \rho \frac{\pi}{4} d^2 C_D \, |\vec{u} - \vec{u}_s| \, (\vec{u} - \vec{u}_s) + \rho \frac{\pi}{6} d^3 C_M \frac{d}{dt} (\vec{u} - \vec{u}_s)$$

$$(6.6.1)$$

where the driving terms on the right hand side are respectively due to gravity, pressure gradients, drag, and added hydrodynamic mass.

This equation has (apart from very short-lived transients) solutions of the form

$$\vec{u}_s = \vec{u} + \vec{w}_o + w_o \, \vec{o}(\varepsilon) \tag{6.6.2}$$

where

$$\varepsilon = \frac{1}{g} \left| \frac{d\vec{u}}{dt} \right| \tag{6.6.3}$$

meaning that, as long as fluid accelerations are small compared with g, the sediment (or bubble) will quickly obtain a velocity which is well described by the simple superposition law:

$$\vec{u}_s = \vec{u} + \vec{w}_o \tag{6.6.4}$$

This simple approximation, combined with the kinematics of the two flows, is sufficient for understanding the differences displayed in Figure 6.6.1. See Nielsen (1992) Section 4.6 for details.

6.6.3 The time scale of particle acceleration

The time scale δ_t for the decay of the transients, i e, the time until (6.6.4) becomes an accurate representation, can be found by considering the case of a particle accelerating from rest towards its terminal settling (rise) velocity w_o in a still fluid. The result, corresponding to Equation (6.6.1) is

$$\delta_t = \frac{s + C_M}{s - 1}\frac{w_o}{g} \tag{6.6.5}$$

which for typical beach sand with $w_o \approx 2.5 \text{cm/s}$ is as little as five milliseconds (taking $C_M = 0.5$ as for a sphere, Liggett (1994), p. 114), and hence without practical consequences.

An estimate of the velocity variance for a particle or bubble can be derived using Nielsen (1992) Equation (4.5.13) which, in the simple case of a simple harmonic, horizontal fluid velocity $u = U\,e^{i\omega t}$ gives the first order approximation for the particle velocity

$$u_p(t) = U_p e^{i\omega t} = \left[1 - \frac{\alpha}{1 - i\dfrac{\alpha g}{w_o \omega}}\right] U\, e^{i\omega t} \tag{6.6.6}$$

where $\alpha = \dfrac{s-1}{s+C_M}$. For heavy particles $(s>1 \sim \alpha>0)$ this indicates $\text{Var}\{u_p\} < \text{Var}\{u\}$. However, for light particles and bubbles we can have $\text{Var}\{u_p\} > \text{Var}\{u\}$ with the limit of $\text{Var}\{u_p\} = 9\text{Var}\{u\}$ for a mass-less sphere $(\alpha \to -2)$ at high frequencies where $g/w_o\omega \ll 1$.

6.6.4 Suspended particles in wave flows

For a particle submerged in a uniform oscillatory velocity field like (approximately) pure wave motion, e g,

$$\bar{u}(x,z,t) = \omega \begin{pmatrix} A\cos \omega t \\ B\sin \omega t \end{pmatrix} \tag{6.6.7}$$

it can similarly be shown that

$$u_s = u(t - \delta_t) + w_o + w_o\, o(\varepsilon^2) \tag{6.6.8}$$

The delay represented by δ_t has no net effect over a wave period and the order ε^2 terms include the settling velocity reduction which occurs if the drag force is non-linear. For a purely vertical oscillatory flow [$A = 0$ in (6.6.7)], and quadratic drag, this settling velocity reduction is given by

249

$$\frac{\overline{w}}{w_o} = 1 - \frac{1}{16}\left(\frac{\omega^2 B}{g}\right)^2 \tag{6.6.9}$$

cf Nielsen (1992), p. 177. Since the acceleration amplitude $\omega^2 B \ll g$ in most natural flows, this non-linear drag correction is not important in praxis.

6.6.5 Suspended particles in vortices

Vortices play an important role for the transport of suspended sediment because they are able to trap sediment particles or bubbles, which have settling (rise) velocity smaller than the typical vortex velocities. This was, most strikingly, pointed out by Tooby et al. (1977), see their photograph in Figure 6.6.2.

The heavy particle on the right and the bubbles on the left all move on (nearly) closed orbits and hence have zero average velocity,

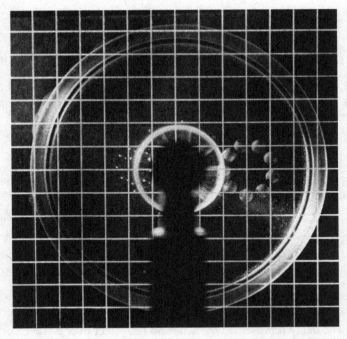

Figure 6.6.2: Stroboscopic photograph, by Tooby et al. (1977), of a heavy particle and some bubbles in a water filled drum. The drum rotates anticlockwise so that water velocities are upwards on the right and downwards on the left.

$$\overline{\vec{u}_s} \equiv \vec{o} \tag{6.6.10}$$

just like the water particles.

In order to end up with zero net velocity, any bubble or particles must sample water velocities with a particular bias. That is, the identity (6.6.9) with (6.6.4) inserted is

$$\overline{\vec{u} + \vec{w}_o}\,\big|_{\text{following a particle}} \equiv \vec{0} \tag{6.6.11}$$

and hence

$$\overline{\vec{u}}_{\text{following a particle}} \equiv -\vec{w}_o \tag{6.6.12}$$

In other words: In a steady vortex flow, both heavy particles and bubbles will move along orbits where the average fluid velocity eliminates their settling (rise) velocity. This is true at the same time as every water particle has $\overline{\vec{u}} \equiv \vec{0}$ and the volume averaged water velocity at any time is also zero: $<\vec{u}(t)> \equiv \vec{o}$.

This truly remarkable result is useful for its direct applications, and for understanding why the inference of particle velocity statistics from the statistics of a surrounding turbulent flow field remains a barrier for researchers. The relevant Lagrangian particle velocity statistics can in praxis only be obtained through particle tracking.

The particle orbits can easily be derived using the simple superposition law (6.6.4): Let the drum be centered at $(x,z) = (0,0)$ and rotate with angular velocity ω. The fluid velocity field is then given by:

$$\vec{u}(x,z) \equiv \omega \begin{pmatrix} -z \\ x \end{pmatrix} \tag{6.6.13}$$

and the particle velocity field, obtained by superposition of the settling velocity is

$$\vec{u}_s(x,z) = \omega \begin{pmatrix} -z \\ x \end{pmatrix} + \begin{pmatrix} 0 \\ -w_o \end{pmatrix} = \omega \begin{pmatrix} -z \\ x - w_o/\omega \end{pmatrix} \tag{6.6.14}$$

The analogy between the two expressions is obvious: when $\vec{u}(x,z)$ corresponds to the circular fluid orbits around the origin, $\vec{u}_s(x,z)$ corresponds to circular orbits around $\begin{pmatrix} 0 \\ w_o/\omega \end{pmatrix}$. The angular velocity of the orbital motion is in both cases ω.

6.6.6   Turbulence effects on settling of particles and rise of bubbles

An intriguing and, until recently unanswered, question is whether turbulence, with zero mean velocity has an effect on the mean settling velocity of particles and/or the rise velocity of bubbles? Is the velocity variance the same for fluid and particles?

Figure 6.6.3:
Would turbulence with zero mean velocity on the average change the settling velocity of particles or the rise velocity of bubbles? Is the velocity variance the same for particles as for fluid?

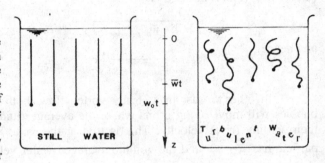

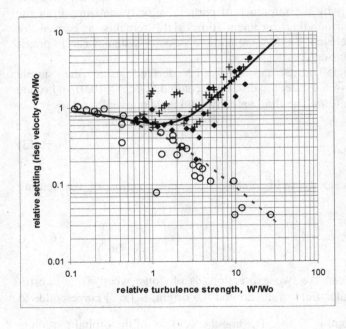

Figure 6.6.4:
Measured settling velocities of dense particles (♦), rise velocities of light particles and bubbles (○), and rise velocities of diesel droplets (+) in turbulence. In these experiments, the turbulence intensity is taken as the rms vertical particle velocity w_s'.

There is now comprehensive experimental data, see Figure 6.6.4, which shows that in fairly weak turbulence, $\sqrt{Var\{w_s\}} = w_s' < 5w_o$, both heavy (solid symbols) and buoyant (open symbols) particles are delayed. This can be understood in terms of the vortex trapping mechanism discussed in connection with Figure 6.6.2.

For very buoyant particles and bubbles, the delay keeps getting stronger with increasing turbulence strength. Surprisingly however, the trend reverses for heavy particles ($\blacklozenge$) around $w_s'/w_o = 3$. In stronger turbulence, the heavy particles speed up again and end up settling at several times their still water rate for $w'/w_o > 10$.

The increased settling rate for heavy particles in strong turbulence can be qualitatively understood in terms of the fast tracking concept of Nielsen (1992, 1993), see Figure 6.6.5.

Nielsen (1992, 1993) also pointed out that this scenario predicts unbounded increase of the settling rate for very strong turbulence:

$$\overline{w_s} \propto w' \quad \text{for} \quad w'/w_o \to \infty \tag{6.6.15}$$

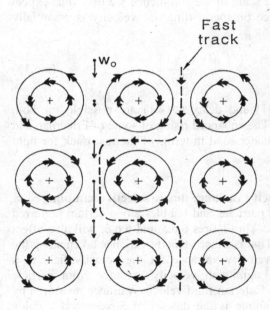

Fast
track

Figure 6.6.5:
In a field of vortices, heavy particles will, by centrifugal effects become concentrated along the vortex boundaries and get swept downwards along the "fast tracks" by the vortex flow. Bubbles and very light particles are however expected to spiral inwards, and get trapped like the bubbles in Figure 6.6.2. A particle following the fast track will experience accelerated settling. It will also experience a reduced velocity variance compared with the flow field as a whole, indeed, the particle sees only downward fluid velocities.

The solid curve, which mimics the trend for heavy particles and diesel droplets in Figure 6.6.4 is given by

$$\frac{\overline{w_s}}{w_o} = \frac{1 + \frac{1}{4}\left(\frac{w'}{w_o}\right)^2}{1 + \frac{w'}{w_o}} \qquad (6.6.16)$$

The dotted curve, which mimics the trend for light particles and bubbles, is given by

$$\frac{\overline{w_s}}{w_o} = \frac{1}{1 + \frac{w'}{w_o}} \qquad (6.6.17)$$

The measurements of Friedman and Katz (2002) on diesel droplets in water showed some dependence of $|\overline{w_s}|/w_o$ upon the Stokes number, St, which is the ratio between the response time scale of the particle and the time scale of the turbulence with the strongest tendency for accelerated rise for particles with $St \approx 1$. Expressing the Stokes number as w_o/gT_L, where T_L is the Lagrangian integral time scale of the turbulence we may thus expect that the influence of turbulence on the settling/rise velocity is essentially given by

$$\frac{\overline{w_s}}{w_o} = \Phi\left(\frac{w'}{w_o}, \frac{w_o g}{T_L}\right) \qquad (6.6.18)$$

where both acceleration ($\Phi > 1$) and delay ($\Phi < 1$) are possible for both buoyant and sinking particles. The enhanced rising of some of Friedman and Katz's diesel droplets can be understood in terms of the fast track for light particles in Figure 6.6.6.

6.6.7 Reduced particle velocity variance due to selective sampling

The concentration of sediment particles and bubbles along certain preferred trajectories, e g, the "fast tracks" in Figures 6.6.5 and 6.6.6, will also affect the particle velocity variance. That is, since velocity variance taken along the "fast track" is less than the velocity variance for the velocity field as a whole, one often finds the Lagrangian velocity variance from particle tracking to be smaller than the Eulerian fluid velocity variance measured by a fixed probe. A classical example is the dataset of Snyder and Lumley

(1971). They found a significant reduction: $\text{Var}\{w_s\} \sim 0.6\text{Var}\{w_{\text{fluid}}\}$ by tracking particles, which were so small ($w_o/gT_L \approx 0.03$), that Equation (6.6.6) predicts only insignificant variance reduction.

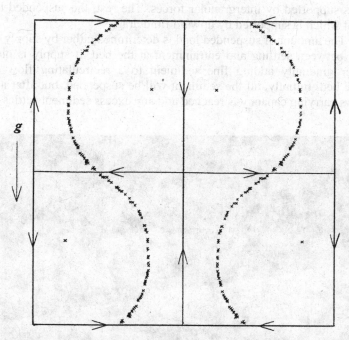

Figure 6.6.6: Pattern of concentrated bubbles in a cellular flow field calculated by Maxey (1990). The bubbles were initially uniformly scattered. The isolated "bubble" in each cell is at the stable neutral point, where $\overrightarrow{u_f} = -\overrightarrow{w_o}$, into which a great number of particles have actually converged. The curves are rising fast tracks which are pieced together from arcs, which within each cell are inward spirals towards the neutral point.

6.7   SEDIMENT SUSPENSIONS

6.7.1   Introduction
In Section 6.4.2, bedload was defined as that part of the sediment load, which is supported by intergranular forces. The rest, the suspended load, is then that which is supported by upward fluid drag.

The amount of suspended load is determined either by supply or by a balance between settling and entrainment at the bed, if supply is plentiful. Consider gradually adding fine sediment to a recirculating flow over a concrete bed. Initially, all the sediment will be suspended, but after a while, the flows carrying capacity is reached and any excess sediment settles out.

Figure 6.7.1: Large breakers bringing suspended sand to the surface. Whale Beach, Sydney Australia.

Sandy sediments are usually carried at capacity, while silt and clay is often carried at loads below capacity. Sediment carried below capacity is sometimes called *wash load*. By its nature wash load is "migratory" rather than "resident" and is therefore only found in rivers and estuaries.

A large part of the sediment transport in surf zones occurs as suspended transport near the breakpoint and in the swash zone. Figure 6.7.1 shows large plunging waves stirring up sand as the plunging jets impinge on the bar. The sand is subsequently brought to the surface by the rising plumes of entrained bubbles, which drive the "fountains" of Figure 6.7.3. After the plumes reach the surface and the air escapes, the sand may travel some distance along the surface before it gradually settles out.

Clouds of suspended sand on the smaller scale of the vortices over vortex ripples was shown in Figure 6.5.6.

The nature of sediment suspensions depends strongly upon the bed geometry. The simplest situation is of course that of a flat bed, where the sediment concentrations can be modelled as functions of elevation and time only: $c = c(z,t)$. Examples of $c(z,t)$-time-series from sheet flow were shown in Figure 6.4.2.

The extensive laboratory experiments by Ribberink et al. (1994, 2000) with $d_{50} = 0.21$mm, 0.32mm and 0.13mm shows that upwards from about 1cm from the undisturbed bed, the corresponding time averaged sediment concentrations vary as $\overline{c}(z) \propto (z/L)^{-2}$, where the vertical scale L is of the order 5mm.

If the bed consists of vortex ripples the vertical scale of the concentration profiles is of the order of the ripple height.

While sediment concentration time series from different levels above a flat bed (e g Figure 6.4.2) show a time variation which is entirely due to the variation of $u_\infty(t)$, the concentration time series from different levels above rippled beds show two or four concentration peaks per wave period, which are due to the passage of sand clouds. These clouds are the remainders of the sediment laden vortices which are released from the lee of the ripples at each flow reversal, cf Figure 6.5.6.

Also over flat or almost flat beds will concentration time series usually show peaks due to the passage of sediment clouds. These clouds will be generated by occasional, more or less randomly spaced vortex shedding and are therefore not periodic. They do not show up in Figure 6.4.2 because they have been averaged out by ensemble averaging over many (regular) wave periods. In field conditions however, where the waves are not periodic and ensemble averaging not possible, these random clouds (the spotted carpet effect) make it very difficult to relate $c(z,t)$ to $u_\infty(t)$.

For both flat and rippled beds the profile shape, $\bar{c}(z)$ may vary considerably depending on the size of sediment considered. An example of this is shown by the data in Figure 6.7.2, which was measured by suction sampling over vortex ripples under waves. Each curve represents one sieve fraction.

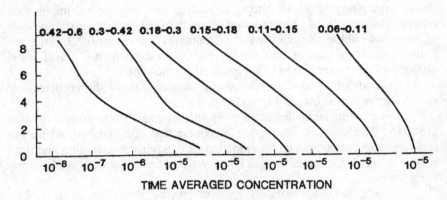

Figure 6.7.2: $\bar{c}(z)$-profiles for different sand fractions in the same flow over a rippled sand bed under non-breaking waves. The numbers on the curves are sand size in mm. The $\bar{c}$-axes are logarithmic. The wave and bedform conditions were: $(h, T, H, d_{50}, \lambda, \eta) = (0.3\text{m}, 1.51\text{s}, 0.13\text{m}, 0.19\text{mm}, 0.078\text{m}, 0.011\text{m})$.

The settling velocity varies by a factor 9 from the finest fraction to the coarsest. The fact that the difference in slope is nowhere near this magnitude indicates strong differential diffusion (Section 6.7.5). The change in shape from upward convex for the fine sand to upward concave for coarser sand is also an interesting modelling challenge.

For the data in Figure 6.7.2 the wave period T is obviously fixed while w_o increases by a factor 9 from the finest to the coarsest sand fraction. Examples with fixed w_o and variable T, i e, same sand in different waves, are shown by Nielsen (1992), p. 217. In that case, similar shape changes occur, which correlate well with the shape parameter $S = w_o T / \sqrt{Ar}$, where r is the hydraulic roughness estimated by (6.5.17). That is, for small S the profiles are upward convex, while for larger S, they become more or less straight $(c \sim e^{-z/L})$ and subsequently upward concave with further increase of S.

258

Wave irregularity leads to smaller bedforms in general and hence to smaller vertical scale of the $\bar{c}(z)$-profiles.

Currents influence the bedforms and add turbulence (compared with non-breaking waves). Hence the $\bar{c}(z)$-profiles are usually stretched vertically due to the presence of a current.

Wave breaking has a strong influence on the $\bar{c}(z)$-profiles because it provides mechanisms for transporting suspended sediment convectively from the bottom to the surface (Figures 6.7.1 and 6.7.3) and because it enhances the turbulent mixing. A set of laboratory data is shown in Figure 6.7.4.

Figure 6.7.3: Fountains of water rising 4–5m in the air after the plunging of swell waves. Figure 6.7.1 offers a view from the back of some of these fountains. In such situations, the plunger jet may impact on the bed and entrain large amounts of sand which is easily carried straight to the surface (and into the air) by the vertical flow velocities which generate the fountain. The vertical velocities $w_{fountain}$ in the "roots" of these fountains can be inferred from the height z_{max} to which the water rises: $w_{fountain} = \sqrt{2gz_{max}}$, in this case around $\sqrt{2 \times 9.8 \times 4} \approx 9 \text{m/s}$.

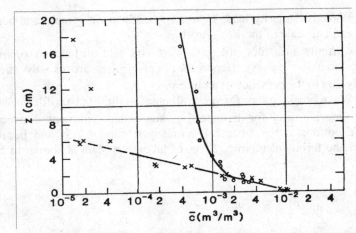

Figure 6.7.4: Suspended sediment concentrations under non-breaking waves (×) and spilling breakers (○) of the same height and period over the same sand bed. T=1.3s, h=0.38m, H=0.19m, d_{50}=0.08mm.

The turbulence, which provides the mixing to keep the sediment in suspension, is to a variable extent dampened by stratification. This effect is visible in the data in Figure 6.7.5.

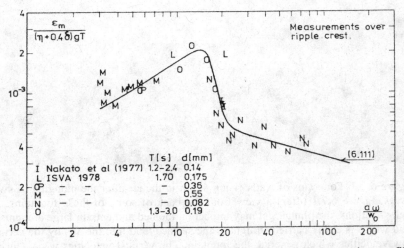

Figure 6.7.5: Relative vertical scale of suspended sediment concentration profiles versus the relative wave velocity $A\omega/w_o$. Rippled beds under non-breaking waves. After Nielsen (1979).

The ordinate is essentially the vertical scale L_c ($\varepsilon_m = w_o L_c$) of the concentration profiles obtained over ripple beds under non-breaking waves.

We see that the concentration scale initially increases with increasing $A\omega/w_o$ but then rather abruptly decreases for $15 < A\omega/w_o < 25$ and then decreases at a slower rate for $A\omega/w_o > 25$. We believe that the reason for the decline is stratification as the concentrations and total sediment loads increase beyond some critical values occurring at about $A\omega/w_o = 15$. The same threshold has recently been obtained from a stratification model for sheet flow by Professor Dan Conley and coworkers, personal communication (2007).

6.7.2 The modelling framework for sediment suspensions
The modelling framework for the suspended sediment concentrations consist of a field equation and boundary conditions at the model boundaries.

The field equation expresses the conservation of sediment

$$\frac{\partial c}{\partial t} = -\nabla \cdot \vec{q}_s = -\nabla \cdot \left\{ \vec{u}_p\left(x,y,z,t\right) c\left(x,y,z,t\right) \right\} \qquad (6.7.1)$$

i e, the local rate of change of sediment concentration equals minus the divergence of the sediment flux field. (The operator "$\nabla \cdot$" is the divergence operator on a vector field $\vec{R} = (R_x, R_y, R_z)$: $\nabla \cdot \vec{R} = \dfrac{\partial R_x}{\partial x} + \dfrac{\partial R_y}{\partial y} + \dfrac{\partial R_z}{\partial z}$).

In the applications it is only the bottom boundary condition that requires detailed attention. It has traditionally been handled in terms of a reference concentration, but this approach is increasingly abandoned in favour of pickup functions, i e, specification of the upward sediment flux rather than the concentration at the base of the modeling domain.

6.7.3 Reference concentrations
Many previous suspension models have relied on the concept of a reference concentration by using a boundary condition of the form $c(z_b, t) = C_b(t)$ at the bottom of the model domain, $z = z_b$.

This approach is however at the most applicable for steady and uniform conditions (even there it may lead to unrealistic results, cf Parker 1978). Data sets like the one in Figure 6.4.2, which show the concentration at the undisturbed bed level to vary less than those 5 or 10 grain diameters above, also show the inappropriateness of $c(0,t)$ as a driver of the sediment concentrations at higher levels. See the discussion of Nielsen et al. (2002) for further details.

6.7.4 Pickup functions

The alternative to a prescribed concentration at the model base is a prescribed upward flux or pickup function $p(t)$, which may plausibly be related to the instantaneous skin friction Shields parameter or for practical purposes to the grain roughness Shields parameter $p(t) = p(\theta_{2.5}[t])$ with $\theta_{2.5}(t)$ given by (6.4.16). For an equilibrium situation the pickup flux must equal the settling flux: $\bar{p} = w_s \overline{c_b}$ but for an unsteady or non-uniform situation a difference may exist, resulting in a sand level change:

$$c_{\max} \frac{dz_s}{dt} = w_s c(0,t) - p(t).$$

Van Rijn (1984) reviewed the pickup function concept for steady flow. He provided new data and a steady flow formula which was adapted by Nielsen (1992, et al. 2002) to oscillatory sheet flow in the form

$$p(t) = \begin{cases} 0.017 w_o \, | \, \theta_{2.5}(t) - 0.05 | & \text{for } \theta_{2.5}(t) > 0.05 \\ 0 & \text{for } \theta_{2.5}(t) < 0.05 \end{cases} \qquad (6.7.2)$$

Nielsen (1979) introduced pickup functions for waves over vortex ripples. For these rippled beds the pickup function has two peaks per wave period, one at each flow reversal where the sand laden lee vortices (Figure 6.5.6) are released into the main flow.

6.7.5 Sediment distribution modelling

The concentration distribution in the interior of the modelling domain is determined by the conservation of sediment:

$$\frac{\partial c}{\partial t} = -\nabla \cdot \vec{q}_s = -\nabla \cdot \left\{ \vec{u}_p (x,y,z,t) \, c(x,y,z,t) \right\} \qquad (6.7.1)$$

where $\vec{u}_p$ is the sediment particle velocity.

This equation may of course be attacked directly with a large enough computer but usually simplification through various kinds of averaging is preferred.

Horizontal averaging is commonly applied over an area which is small compared to the scales of wave deformation and beach morphology but greater than the scale of ripples or of turbulent bursts over a flat bed.

Time averaging of c and u separately is usually not recommendable because the transport contribution from $\tilde{u}_p \tilde{c}$ is quite often greater than that from $\bar{u}_p \bar{c}$. Thus, the periodic component $\tilde{c}$ must be calculated as well as $\bar{c}$.

In order to get started on modelling, a horizontally uniform situation is often assumed. This leads to the following, simplified version of (6.7.1):

$$\frac{\partial c}{\partial t} = -\frac{\partial q_{sz}}{\partial z} = -\frac{\partial}{\partial z}\{w_p(z,t)\,c(z,t)\} \approx -\frac{\partial}{\partial z}\{w(z,t)c(z,t)-w_s c(z,t)\}$$

(6.7.3)

where the issues of time dependence may be illustrated by considering a periodic flow where $w = 0 + \tilde{w} + w'$ and $c = \bar{c} + \tilde{c} + c'$ ($\bar{w} \equiv 0$ for a horizontally uniform flow). From phase averaging we then get

$$\frac{\partial \tilde{c}}{\partial t} = -\frac{\partial}{\partial z}\{\tilde{w}\bar{c} + \tilde{w}\tilde{c} + w'c' - w_s\tilde{c}\} \qquad (6.7.4)$$

and, by further averaging over a few wave periods

$$0 = \frac{\partial}{\partial z}\{\overline{\tilde{w}\tilde{c}} + \overline{w'c'} - w_s\bar{c}\} \qquad (6.7.5)$$

These equations have traditionally been dealt with in terms of gradient diffusion, e g, $\overline{\tilde{w}\tilde{c}} + \overline{w'c'} = -K_{\text{Fick}}\dfrac{\partial \bar{c}}{\partial z}$ for (6.7.5). That is however an oversimplification, which misses the differential diffusion displayed in Figure 6.7.2 and by several other classical data sets. A more satisfactory representation is obtained with a finite mixing length model as shown below.

6.7.5.1 The mixing length approach

Rather than trying to account for all the details of $w = \bar{w} + \tilde{w} + w'$ and $c = \bar{c} + \tilde{c} + c'$ it is usually more appropriate to consider the sediment flux in terms of the mixing flux q_m described in Section 5.4.7 and the settling flux $w_s c$: $q_{sz} = q_m - w_s c$. The conservation equation (6.7.3) then becomes

$$\frac{\partial c}{\partial t} = -\frac{\partial}{\partial z}(q_m - w_s c) = -\frac{\partial}{\partial z}\left[w_m[c(z-\frac{l_m}{2}) - c(z+\frac{l_m}{2})] - w_s c(z)\right] \quad (6.7.6)$$

263

Sediment transport mechanisms

which by time averaging (permitted when the wave period is much longer than the turbulent time scale) becomes

$$0 = \frac{\partial}{\partial z}\left[-w_m[\overline{c}(z-\frac{l_m}{2})-\overline{c}(z+\frac{l_m}{2})]-w_s\overline{c}(z)\right] \qquad (6.7.7)$$

Further simplification is achieved by assuming that the [ ], which represents the total vertical flux of sediment, is zero at the free surface and hence is zero everywhere as $\partial/\partial z=0$:

$$w_m[\overline{c}(z-\frac{l_m}{2})-\overline{c}(z+\frac{l_m}{2})]-w_s\overline{c}(z) = 0 \qquad (5.4.35)$$

In Section 5.4.7 this equation was solved for the case of homogeneous turbulence, i e, constant w_m, l_m. It was found that the solution to this complete equation is analogous to that of the Fickian approximation (obtained by assuming $\overline{c}(z-\frac{l_m}{2})-\overline{c}(z+\frac{l_m}{2}) = -l_m\frac{\partial\overline{c}}{\partial z}$), but that only the complete equation explains differential diffusion. That is, coarser sediment (greater w_s) has a greater apparent Fickian diffusivity than fine sediment:

$$K_{\text{Fick}}(w_s) = w_m l_m\left[1+\frac{1}{24}(\frac{w_s}{w_m})^2+...\right] \qquad (5.4.39)$$

which is indeed the trend shown by experimental data where the concentration profiles of different sediment fractions have been measured in the same flow, cf Figure 6.7.2 for oscillatory flows and Coleman (1970) for steady flow. In other words, a simple gradient diffusion approach with a sediment diffusivity, which is independent of w_s is inadequate.

Van Rijn (1984b) suggested an ad hoc approach expressed as $K_{Fick}(w_s) = \beta v_t = v_t\left[1+2(\frac{w_s}{u_*})^2\right]$ to mimic experimental findings in line with (5.4.39) if $w_m \sim u_*$. This ad hoc fix may be a reasonable engineering approach as long as it is realised that *Reynolds hypothesis*: $K_{Fick}= v_t$ does not hold for suspended sediment. It does however not enable modelling the different profile shapes displayed in Figure 6.7.2.

We shall see below that similar results to (5.4.39) are obtained with various different pairs (w_m, l_m), which are relevant to natural flows, and that the details of Figure 6.7.2 are, at least qualitatively, captured by the mixing length approach with appropriate choices of $l_m(z)$ and $w_m(z)$.

6.7.5.2 Mixing length model of $u(z)$ and $c(z)$ for $l_m = \lambda z$ and $w_m = \gamma u_*$

First consider the flow near the bed in a river or under an ocean current. That is, a flow which has traditionally been modelled as a constant shear stress layer ($\tau \approx \tau_o = \rho u_*^2$) with a linearly increasing eddy viscosity, $v_t = \kappa u_* z$. (This corresponds to the bottom part of the logarithmic velocity profile discussed in Section 6.2.2). More precisely, we consider a flow where the mixing velocity is constant $w_m = \gamma u_*$, while the mixing length grows linearly with distance from the bed: $l_m = \lambda z$.

First we find the parameter space (λ, γ), which corresponds to the experimentally verified logarithmic velocity distribution: Writing the upward flux of x-momentum ($-\tau = -\rho u_*^2$) as a mixing flux in the form of Equation (5.4.28) with $w_m = \gamma u_*$, and $l_m = \lambda z$, we get

$$-\rho u_*^2 = \gamma u_* \left(\rho u \left\{ z - \frac{1}{2} \lambda z \right\} - \rho u \left\{ z + \frac{1}{2} \lambda z \right\} \right) \qquad (6.7.8)$$

where insertion of the log-law

$$u(z) = \frac{u_*}{\kappa} \ln \frac{z}{z_o} \qquad (6.2.4)$$

gives

$$\gamma \ln \frac{1 + \lambda / 2}{1 - \lambda / 2} = \kappa \qquad (6.7.9)$$

This equation, together with the physical limits $0 < \lambda < 2$ (cf Figure 5.4.9), defines our limits of choice for (λ, γ).

Turning now to the sediment, we have to solve (5.4.39) with $w_m = \gamma u_*$, and $l_m = \lambda z$, i e,

$$\gamma u_* \left(\overline{c} \left\{ z - \frac{1}{2} \lambda z \right\} - \overline{c} \left\{ z + \frac{1}{2} \lambda z \right\} \right) - w_s \overline{c}(z) = 0 \qquad (6.7.10)$$

As for homogeneous turbulence (Section 5.4.8) the solution to this complete, finite-mixing-length equation, is similar to that of the Fickian, infinitesimal-mixing-length (i.m.l) approximation. The i.m.l approximation

265

$$-\gamma u_* \lambda z \frac{\partial \overline{c}}{\partial z} - w_s \overline{c} = 0 \qquad (6.7.11)$$

corresponding to $\overline{c}\left\{z - \frac{1}{2}\lambda z\right\} - \overline{c}\left\{z + \frac{1}{2}\lambda z\right\} = -\lambda z \frac{\partial \overline{c}}{\partial z}$, has the solution

$$\overline{c}(z) = \overline{c}(z_o)\left(\frac{z}{z_o}\right)^{-\frac{w_s}{\lambda \gamma u_*}} \qquad (6.7.12)$$

which corresponds to a classical i.m.l model by Rouse (1937) for $\lambda \gamma = \kappa$.

Inserting an analogous power function expression, e g, $c(z) = c(z_0)(z/z_0)^p$ into (6.7.10), leads to the following equation for the power p

$$(1 - \frac{\lambda}{2})^p - (1 + \frac{\lambda}{2})^p = \frac{w_s}{\gamma u_*} \qquad (6.7.13)$$

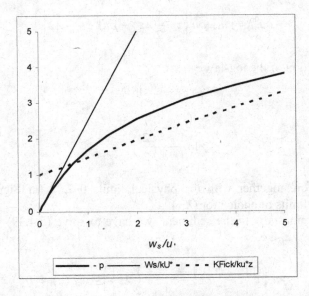

Figure 6.7.6: The exponent p compared with the "infinitesimal mixing length" value $-w_s/\kappa u_*$ for $(\lambda,\gamma) = (1, 0.36)$. The finite mixing length behaviour quantified by w_s/ku_*-p, is stronger for larger λ. Also shown is $\beta = K_{\text{Fick}}/v_t = K_{\text{Fick}}/\kappa u_* z$ where $-K_{\text{Fick}} \frac{\partial c}{\partial z} - w_s c = 0 \iff K_{\text{Fick}} = \frac{-w_s z}{p}$.

The behaviour of p according to this equation is illustrated by Figure 6.7.6. We see that the finite mixing length effects are negligible for very fine sediment, i e, $p \rightarrow -w_s/\kappa u_*$ and $\beta = K_{Fick}/v_t \rightarrow 1$ for $w_s/u_* \rightarrow 0$. The increasing trend for $\beta = \beta(w_s/u_*)$ is in good qualitative agreement with flume studies, e g Coleman (1970) and Graf and Cellino (2002). However, some details are still not understood, including the occurrence of $\beta<1$ for some flat bed experiments by Coleman (1970).

6.7.5.3 Suspension under waves
Application of Equation (5.4.35) with $l_m = \lambda z$ and $w_m = \gamma u_* \exp(-z/L_w)$ can also explain the qualitatively different $\overline{c}(z)$-profiles for fine and coarse suspended sediment over wave ripples, cf Figure 6.7.2. For full details see Teakle and Nielsen (2004). The traditional Fickian (gradient diffusion) approach could never explain the different shapes of the $c(z)$-profiles displayed by fine and coarse sand in a given flow of this type.

Under breaking waves or in combined wave current flows, where the turbulence is stronger at greater distance from the bed (cf Figure 6.7.3), the model should probably be modified to $l_m = \lambda z$ and $w_m = \gamma u_* /(1+z/L_w)$.

6.7.6   Suspensions of cohesive sediments
Suspensions of cohesive sediments are more complicated because the effective settling velocity varies due to flocculation, which depends on salinity and turbulence intensity as described by Winterwerp (1998).

Salinity enhances flocculation by cushioning electrically charged particles against the electrostatic repulsion, allowing the van der Waal's forces to bind the particles together. Hence, fine sediment which encounters salt water seaward of the null point of the salt wedge, Figure 5.3.5, will sink and return upstream with the intruding saltwater. There is hence a convergence of fine sediment near the null point which gives rise to the frequently observed turbidity maxima in this area. Open University (1999), p. 164 shows details of the turbidity maximum in the Seine estuary and Ellis et al. (2008) present a recent model of turbidity maxima in the Irish Sea.

6.8   BASIC SEDIMENT TRANSPORT MODEL BUILDING

6.8.1   Introduction

The aim of the following is, where possible, to deliver the building blocks for morphological change models. The task is easy in the case of sheet flow under a pure wave motion where the Meyer-Peter and Müller type formulae (6.4.16–6.4.18) may be applied. However, when considerable amounts of sediment moves in suspension, the timing of entrainment and settling must be given detailed consideration.

When currents are present they may have quite different directions and vertical distributions from the wave induced velocities. Many of these issues are still beyond the state of the art, and most of the data, available for model verification is 2DV only.

6.8.2   Two different families of transport models

Models of the short term averaged sediment transport rate come in two distinct varieties: The traditional $u \times c$-integral models with the general form $\overrightarrow{q_s} = \int_{bed}^{surface} c\,\overrightarrow{u_s}\,dz$ and the much less common particle trajectory models which are based on the pickup function concept and the average trajectory of moving sediment particles, Figure 6.8.1.

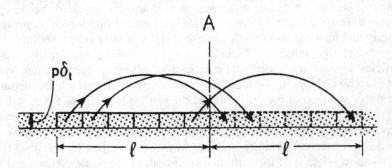

Figure 6.8.1: The amount of sand picked up per unit time is the pickup function $p(t)$ and on the average $\overline{p}$. Then, if the average jump length is l, the flux density through a vertical cross section (A) is seen to be $\overline{q_s} = \overline{p}l$ (where l and then q may be vectors in the horizontal plane).

The literature contains a vast number of $u \times c$ -integral models and a recent guide to the building of such models can be found in van Rijn (2007 I–IV). One important general consideration about these models is that the net transport due to oscillatory components is not to be ignored, i e, the term $\overline{\tilde{u}_s \tilde{c}}$ is often at least as important as $\overline{u_s} \, \overline{c}$.

A very simple and very successful model of the particle trajectory type for shore normal transport over vortex ripples was developed by Nielsen (1988) and shown to be more reliable across different sand sizes than more complicated models of the $u \times c$-integral type.

6.9   SEDIMENT TRANSPORT OUTSIDE THE SURF ZONE

6.9.1   Shorenormal transport
Outside the surf zone, the experience from detailed large scale laboratory studies with non-breaking waves or oscillating water tunnels applies. Hence, Meyer-Peter and Müller type transport formulae (6.4.15)–(6.4.18) are fairly reliable for sheet flow conditions.

For rippled beds, the accuracy of available models is not as good. The trouble starts with the considerable uncertainty in predicting the ripple geometry. This has impact on the estimation of near-bed concentration magnitude (higher, sharper ripples → greater stresses at the crest → larger concentrations) as well as on the distribution of suspended sediment concentrations and fluid velocities. For a given ripple geometry, the finer details of the suspended sediment concentrations, mentioned in connection with Figure 6.7.2 are only modelled with computationally expensive models and the modelling of the stratification effects, clearly visible in Figure 6.7.5, is currently an area of new research. Simple models, corresponding to conventional diffusion modelling and two alternative approaches, were developed by the writer in the 1980s and tested against small scale laboratory data, cf Nielsen (1992) p. 266. These models can be calibrated for field use, given suitable large scale measurements.

6.9.2   Shoreparallel transport outside the surf zone
The shoreparallel sediment transport outside the surf zone is usually insignificant or dominated by strong currents of tidal or oceanic origin. Cherlet et al. (2007) is a recent example of modelling this sediment transport and the resulting bedform dynamics.

6.10 SURF ZONE SEDIMENT TRANSPORT

6.10.1 Shorenormal transport

The modelling of shore normal surf zone sediment transport q_x in a purely 2D situation, such as a wave flume, has progressed to a stage where the resulting beach profile developments have been modelled with reasonable accuracy, at least for irregular waves where the morphological features are not too sharp, see e g, Roelvink and Broeker (1993). Some of these good end results are however somewhat fortuitous in that none of the models deal with acceleration effects or streaming which have since been shown to be very important, see Section 6.4.8. Other challenges are in the details of the flow near the breakpoint and in the distribution of the undertow which may or may not overpower the bottom boundary layer streaming.

The suspended load is strongly influenced by wave breaking as indicated by Figure 6.7.4, and the breaking induced mixing is highly variable between areas of strong organized upward motion, Figures 6.7.1 and 6.7.3, and gently spilling breakers.

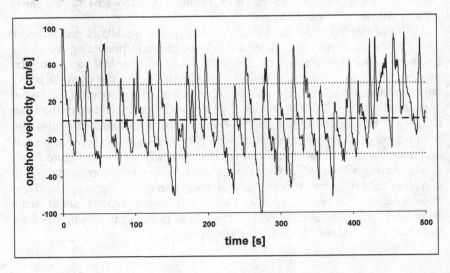

Figure 6.10.1: An example of nearbed (z~20cm) shorenormal velocity time series in the surf zone. $\bar{u} = 0$, stdev$\{u\} = 38$cm/s, $\overline{u^3} / \left(\overline{u^2}\right)^{3/2} = 0.23$, $\overline{\left(\dfrac{du}{dt}\right)^3} / \left(\overline{\left(\dfrac{du}{dt}\right)^2}\right)^{3/2} = 1.30$.

Neglecting the unsolved problems with the suspended load, i e, just trying to model the transport closest to the bed with a Meyer-Peter and Müller type transport formula like (6.4.15)–(6.4.18) is not even sure to succeed. The problem is that wave shape is so variable in the surf zone, Figure 6.10.1, and that surf beat is usually present.

Application of a single typical wave shape like the one parameterized by Elfrink et al. (2006) may then give quite meaningless sediment transport rates due to the highly non-linear sediment transport process. Comprehensive calibration with detailed large scale data is needed.

Practical beach erosion models still often use the simple approach of Kriebel and Dean (1985), where the offshore sediment transport rate is assumed proportional to the excess energy dissipation rate per unit volume compared with equilibrium conditions.

6.10.2 Shoreparallel transport

For the shoreparallel sediment transport $q_{sy}(x)$ in the surf zone, there are now also a number of detailed models of the $u \times c$-integral type mentioned above. They are however still not very successful with respect to modelling the distribution of the sediment transport across the surf zone, cf Bayram et al. (2001). It is of course not expected, that the distribution of the longshore sediment transport can be modelled with confidence, when the distribution of the longshore current still cannot, see Section 2.5.4. Even if the longshore currents could be modelled, there are however many unresolved questions with respect to the sediment entrainment and bedform geometry across the surf zone.

A reasonable practical alternative is still the empirical formulae for the overall transport rate Q_{sy}. For example the so-called CERC formula

$$Q_{sy} = \frac{K}{16(s-1)\sqrt{\gamma_b}} \sqrt{g} H_b^{2.5} \sin 2\alpha_b \qquad (6.10.1)$$

where γ_b is the breaker index H_b/h_b. This formula is remarkable in that, it does not contain the grain size. This is counterintuitive, but reasonably supported by data within the most common range of grain sizes: 0.18mm$<$ $d_{50}$$<$0.6mm. For coarser sediment, the transport rate tends to be a decreasing function of grain size as indicated by the data for K in Figure 6.10.2, but so far the field data does not define a clear trend.

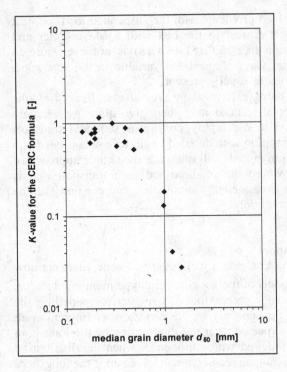

Figure 6.10.2:
Field experimental values of the K coefficient in the CERC formula. Data from Del Valle et al. (1993).

An alternative formula proposed by Kamphuis (1990) on the basis of both laboratory and field data has only very weak grain size dependence:

$$\frac{Q_{sy}}{H_{b,rms}^3 / T_p} = 2.6 \times 10^{-3} \left(\frac{H_{b,rms}}{L_{o,p}} \right)^{-1.25} m_b^{0.75} \left(\frac{H_{b,rms}}{d_{50}} \right)^{0.25} \left(\sin 2\alpha_b \right)^{0.6} \quad (6.10.2)$$

where m_b is the beach slope near the breakpoint. Note that the Kamphuis formula, which is based on data spanning a larger range of wave heights, has a weaker wave height dependence: $Q_{sy} \sim H^2$ compared with $Q_{sy} \sim H^{2.5}$ for the CERC formula.

6.11 SWASH ZONE SEDIMENT TRANSPORT

Sediment transport in the swash zone is an area of great challenge for present and future research. It is obvious that all the sediments which move onshore and eventually on to the coastal dunes on accreting beaches must be transported through the swash zone, yet we do not understand the details of how it happens. Similarly, we lack understanding of the details of longshore transport in the swash zone although this has long been recognized as one of the major contributing areas together with the area around the breakpoint, see e g, Bayram et al. (2001). The challenges lie in all areas from the overall water motion through the boundary layer flow structure to the sediment dynamics. A special concern with respect to the swash zone sediment motion is the possible influence of in- or outflow of water through sand surface, see Section 6.3.6. It is also possible, that the horizontal pressure gradients in the bed under a swash bore fronts (cf Figure 2.4.1) are strong enough to cause fluidization, which enhances the sediment transport rate. For a review of recent developments see, e g, Weir et al. (2006) and Guard and Baldock (2007).

Additional complications are related to surf beat and the associated, occasionally very violent backwash, see Figure 2.3.6. These events are undoubtedly very significant contributors to the erosion during storms.

At the very top of the swash zone conditions are qualitatively different depending on whether the slope is continuous or ends at the foot of an erosion scarp, cf Figure 7.8. The process of scarp retreat on an eroding beach is yet to given a sensible physical description.

6.11.1 Swash sediment transport and the beach watertable

The relations between beach watertable heights and beach erosion have been the subject of much research in recent years. It is obvious that if the beach is very "thirsty", some sand will get deposited near the runup limit because the water drains away around it. It may also be the case that a lower watertable and hence relatively more in- than ex-filtration through the beach face makes the beach more resilient to erosion. The forces on sediment particles corresponding to in-/ex-filtration are discussed in Section 6.3.6.

The potential for beach stabilization through watertable manipulation has prompted the emergence of several commercial systems with claimed benefit in terms of erosion protection. Turner and Leatherman (1997) reviewed the performance of pumped dewatering systems deployed in the field. They found scant evidence of positive effects and several pumping systems have been lost during storms on open coasts. The writers experience with pumped dewatering systems at smaller scales is that spectacular effects can be obtained at small laboratory scales $h\sim10$cm with strong pumps, while the effects are difficult to identify experimentally already at $h\sim50$cm.

Due to the considerable cost of pumping systems and inspired by the fact that the watertable in the beach is higher than the MWS of the inner surf zone, as explained in Section 8.2, a pump-free dewatering system was tested at Dee Why Beach near Sydney in 1991, and the results were reported in detail by Davis et al. 1992, see Figure 6.11.1.

Figure 6.11.1a: Beach drains made like 900mm wide egg cartons wrapped in filter cloth ready for deployment on Dee Why Beach, Sydney.

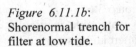 

Figure 6.11.1b: Shorenormal trench for filter at low tide.

Figure 6.11.1c:
Outflow of beach groundwater from the bottom end of one of the drains at low tide.

Figure 6.11.1d:
The presence of two of the drains is clearly shown by tongues of dryer sand at mid tide.

The pump free dewatering system shown above had a clearly measurable effect on the beach watertable, which is visible via the seaward extension of the dry-looking beach face above the drains in Figure 6.11.1d. There were however no significant morphological effects. That is, any morphological effect of the drains was overpowered by the natural signal due to rip system migration etc.

6.12   CONCLUDING REMARKS ON SEDIMENT TRANSPORT MODELLING

The state of the art of hydrodynamics and sediment transport modelling is still not progressed to the level where the development of complex topography can be predicted. Such capabilities may be achievable in a few years for simple 2D beach profiles, but the morphodynamics of 3D sand banks and deltas lies many years ahead.

The most reliable predictive tool is thus still extrapolation of historical data where the main climatic and geological conditions have not changed.

Where these overall defining circumstances are changing one should use a combination of physics based models and intuition based on the study of morphodynamics in nature and in cyberspace.

7 Morphodynamics

7.1   INTRODUCTION

Coastal and estuarine morphologies result from the interplay of hydrodynamic forcing and geological settings. The relative importance of these two depends on how strong the forcing is. Where the forcing is very strong, as on exposed coasts, it is thus fairly easy to interpret the morphodynamics as consequences of the forcing. We shall attempt this in the following, while the reader is referred to collections of case studies like Nordstrom and Roman (1996) for the morphodynamics of the less energetic estuarine shores and deltas.

The coastlines of the world can initially be divided in accordance with the dominant materials, i e, rocky, sandy and muddy shores. Alternatively they could be categorised as either eroding or accreting.

Some shores made of durable rock have changed little since the last ice age where loose sediments may have been removed down to a considerable depth below the present sea level, e g, the fjords of Norway and New Zealand. Less durable rocky coasts e g, the Sydney Sandstone in Figure 7.1.1, and limestone cliffs, Figure 7.1.2, may change considerably on the time scale of centuries or even decades.

Figure 7.1.1: Eroding sand stone headland with the typical rock platform near MSL.

277

Figure 7.1.2:
Cretaceous limestone, which was pushed up to heights of the order 100m by glaciers during the last ice age, are now being eroded at quite a fast rate. The soft chalk is quickly carried away, while flint stones are left behind to form cobble beaches. Møns Klint, Denmark.

The sandy coasts which represent the majority of the world's shores are much more dynamic, particularly near large deltas with ample supply of river sediments and with exposure to large waves which will sort and re-distribute the material.

Areas with large supply of fine river sediments and protection from large waves develop muddy or silty coastlines, for example, the Yellow Sea coast of China, the Caribbean coast of South America and most of the coast of New Guinea.

There is a large literature on coastal geomorphology of which Masselink and Hughes (2003), Short (1999), Komar (1999) and Dean and Dalrymple (2002) are recent examples, which also include summaries of previous works.

Because they offer many engineering challenges as well as the greatest tourist amenities, sandy shores have attracted the most attention. They are also the main subject of the following sections.

Sandy coasts may suffer long term erosion if the sand supply is diminished or totally cut of. However, it must also be realised that many sandy coastlines go through natural cycles with shoreline movements through tens of metres in a few decades. The property damage in Figure

Figure 7.1.3: Houses lost to rapid erosion of coastal dunes. Central Coast New South Wales.

7.1.3 is a result of this natural variability being ignored when the houses were built.

Onshore winds of sufficient strength are able to transport sand landwards to form foredunes (Figure 7.2.3) and in some cases large migrating dune systems. Mt. Tempest on Moreton Island near Brisbane, which rises to 280m, was formed by wind-blown beach sand.

Figure 7.1.4: Landward (left to right) migrating dunes on Moreton Island. Young dunes move into a swamp left after the old dune has moved on.

279

Perhaps contrary to general belief, a lot of aeolean sand transport occurs during rain, cf Hotta et al. (1984). While moist sand can be held down by surface tension in the absence of rain, impacting raindrops eliminate the surface tension and lift some of the sand into the air for transportation by the wind. Persons who walk barelegged on a sandy beach on a windy, rainy day experience the stinging of the wind blown sand hitting the skin.

Figure 7.1.5: Migrating dunes sometimes endanger buildings. Rubjerg Knude, Danish west coast.

Some coastal hydrodynamic processes are able to sort the beach material with respect to density or grain size. The sorting of chalk powder from flint cobbles in Figure 7.1.2 is well understood while the sorting of heavy versus light sand in Figure 7.1.6 is not.

Under erosive storm conditions the heavy minerals are left as a lag deposit in front of the scarp, Figure 7.1.6. The lower plateau was formed during the last high tide of the storm, while the upper scarp was active during the second last high tide. Under accreting conditions the minerals tend to get mixed fairly uniformly.

Figure 7.1.6: Dark, heavy ($s \approx 4.2$) minerals alternating with quartz ($s \approx 2.65$) in front of an erosion scarp, Bribie Island North of Brisbane, Australia.

7.2   BEACH PROFILES

Keeping in mind that most beaches are three-dimensional, the concept of "The beach profile" is still a useful concept in the description of coastal geomorphology. The width of the active profile is only clearly defined in relation to a specific time scale e g, the tidal cycle, the typical lifetime of buildings or the typical length of a glaciation cycle.

For coastal management purposes, the relevant time scale is the typical lifetime of coastal structures, i e, a few decades. The corresponding width of the active profile is then from the seaward convergence of survey profiles at depth h_{cl}, the depth of closure, to the permanent land vegetation. Both of these may change rapidly if the system is knocked out of equilibrium by sea level rise or disruption of the sand supply.

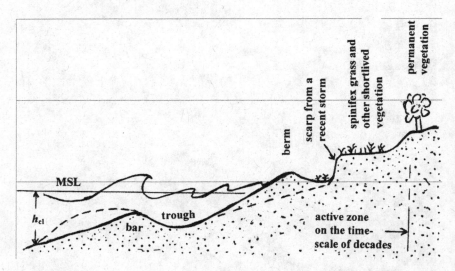

Figure 7.2.1: Definitions and terminology. The dotted line indicates the profile during a recent storm when the scarp was formed and the bar extended further seaward. Some beaches have several bars and beaches with large tidal range may have a clearly defined steep upper profile of coarse material and a flat lower profile of finer material.

7.2.1   "Equilibrium profile shape"

The *"Equilibrium profile"* is one of the oldest concepts in coastal geomorphology. The idea is that, for a given set of wave, tide and sediment parameters there exists an equilibrium beach profile shape towards which the beach will move irrespective of the starting geometry.

This nice simple concept may however not exist. The geometry of a beach profile exposed to fixed wave parameters does not necessarily converge. Instead, there often seems to be an asymptotic cycle. The useful concept that does exist however is the *expected beach profile* given a certain set of wave, tide and sediment parameters.

The most important single parameter, determining profile shape as function of wave and sediment parameters is the *Gourlay parameter* H_b/w_oT (Gourlay 1968), and the influence of the tidal range can be accounted for in terms of the ratio between tidal range and wave height, i e,

$$beach\ profile\ shape\ = \Phi\left(\frac{H_b}{w_oT}, \frac{Tide\ range}{H_b}\right) \qquad (7.2.1)$$

The most commonly assumed shape for the "equilibrium profile" is

$$depth\ (to\ MSL)\ = x^{2/3}\ A\left(\frac{H_b}{w_oT}\right) \qquad (7.2.2)$$

where x is the distance from the MSL shoreline. The 2/3 power can be arrived at theoretically either by seeking constant bed shear stress magnitude (Bruun 1954), or constant energy dissipation per unit volume (Dean 1991).

Dean (1991) found a relation shown between $A[m^{1/3}]$ and the grain size for field conditions ranging from beaches of fine sand (0.1mm) to failed rock breakwaters ($d_{50}>1m$) on the East coast of the USA. However, considering the general trend of decreasing beach slope with increasing H/w_sT in Figure 7.2.2 one would expect that a better predictor could be provided in terms of this parameter or other combinations of sediment and typical wave parameters.

The shape given by (7.2.2) is a useful practical approximation although it does not model bars and gives an unrealistic, infinite slope at $x = 0$. The alternate approach of using an exponential function, i e, $h(x) = B\ [1 - exp\ (-x/L_x)]$ was suggested by Bodge (1992). He found good fitting potential but did not analyse the relation between the depth and length scales (B, L_x) and sediment and wave parameters.

7.2.2   Profile shape variability

The detailed survey study by AD Short (Short 1999) gives a good impression of the temporal variability of the profile shape for micro-tidal beaches. See Figure 7.2.2.

It is interesting and somewhat surprising to note that the most dramatic profile variability is observed for beaches with medium (not large) modal values of H_b/w_oT. Surprising because relatively large H_b/w_oT might have been expected to give the largest sediment mobility. This fact is related to the fact that littoral drift data show very little grain size dependence (in agreement with Equation (6.10.1)).

The profiles from Mid Seven Mile and Eastern Beach indicate a fairly constant volume of sand. This reflects the fact that these beaches are close to the longshore uniform, dissipative extreme Figure 2.1.1 top. The Narrabeen transect, on the other hand, shows great changes in sand volume. This is caused by the highly three-dimensional topography of this beach, often with rip cells like the beach in Figure 2.6.1. Migration of rip cells through the survey transect will appear as large variations in profile shape and sand volume.

Above the normal high tide shoreline, the cyclic behaviour of beaches during and between storms is visible: Around high tide during storms the large waves may create an erosion scarp, which can be several metres high, Figure 7.2.3 top.

After the storm, the scarp becomes rounded and usually covered by vegetation, and a foredune with new vegetation develops in front, Figure 7.2.3 bottom. The foredune may eventually completely cover the scarp.

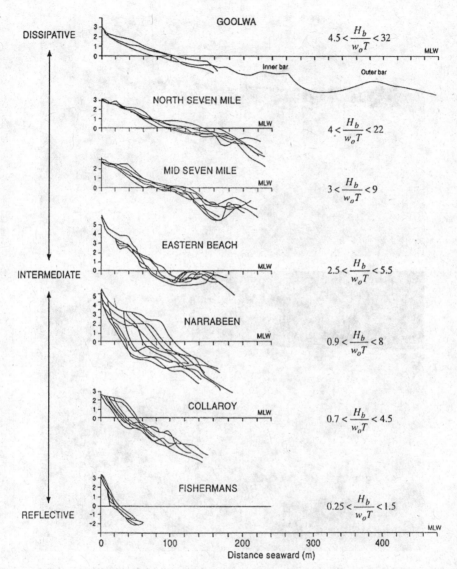

Figure 7.2.2: Successive profiles of a range of beaches with different ranges of H_b/w_oT. After Short (1999). All of these beaches are micro-tidal, i e, the typical tidal range is smaller than the typical significant wave height.

Figure 7.2.3 Top: Fresh erosion scarp formed around high tide (Kingscliff, New South Wales) and *Below:* Foredunes with pioneering vegetation forming in front of an old scarp (Moreton Island).

7.2.3   Tide range influence on beach morphology

While the waves are the dominant sediment stirrers and generators of shore normal currents, the tide's contributions to morphodynamics is to shift the still water surface up and down once or twice a day and sometimes generate longshore currents. This increases the height of the active beach profile. Masselink and Hughes (2003) give a comprehensive description of typical beach morphology as function of relative wave height and relative tide range, i e of $\left(\dfrac{H}{w_s T}, \dfrac{\text{Tidal range}}{H} \right)$.

Macro-tidal beaches often look very different at high and low tide. The high tide beach is then a steep sandy beach while the low tide beach is a mud/silt flat as in Figure 7.2.4.

Figure 7.2.4: Nudge Beach Brisbane, a macro-tidal beach with a steep, sandy high tide beach, and a silty low tide terrace. The slope break is just above MHWN. Photo by Paul Guard.

Figure 7.2.5: Salt pan sequence near Eprapah Creek Brisbane.

Where the wave action is insignificant compared with the tide, the shore may get lined with mangroves within the neap tidal range, i e, between MLWN and MHWN (lower left of Figure 7.2.5).

This bedlevel range is inundated every tidal cycle and the groundwater salinity is approximately that of sea water. Between MHWN and MHWS inundation is less frequent and hypersaline conditions (central bare part of Figure 7.2.5) occur after a few days of evaporation. In this band only plants which are especially salt tolerant can grow.

Upwards from about MHWS rainfall dominates over seawater inundation and salinity becomes lower again allowing for normal terrestrial vegetation, upper left of Figure 7.2.5.

7.2.4   Depth of closure
The depth of closure h_{cl} is a concept of practical importance in connection with sediment budgets and beach nourishment. It is the depth at the seaward limit of the active beach profile and is defined in practical terms as the depth

at which seasonal bedlevel changes become smaller than the survey accuracy, see e g, Birkemeier (1985).

Hallermeier (1981) gave a widely used expression for h_{cl} in terms of the annual average and standard deviation of the significant wave height $\overline{H}_{\text{sig}}$, $\sigma_{H_{\text{sig}}}$ and the corresponding period T_{sig}:

$$h_{cl} = \left[\overline{H}_{\text{sig}} - 0.3\sigma_{H_{\text{sig}}}\right] T_{\text{sig}} \sqrt{\frac{g}{5000 d_{50}}} \qquad (7.2.3)$$

7.2.5 Storm erosion and time scales of profile change

Beach erosion and inundation during storms is one of the greatest concerns in coastal management, particularly in places like The Netherlands where much of the inhabited land is close to or even below the sea level. Consequently, numerous studies have been carried out on the erosion of beaches during storms, e g, the laboratory studies of van de Graaff (1977), Vellinga (1986) and van Rijn (2009).

Storm erosion of natural beaches is illustrated by the scarp-foredune-cycle of Figure 7.2.3 and by the beach profiles in Figure 7.2.2. In a storm, which follows a good long period of fair weather, the beaches will go from the steepest overall profiles to the flattest with a lot of sand from the pre-storm berm moving out into the breakpoint bar, Figure 7.2.1.

For a given storm wave height and duration, the amount of shoreline retreat and the corresponding erosion volume above MSL is greater when

- the runup scale $L_R \sim \sqrt{HL_o} \sim T\sqrt{H}$ of the storm waves is greater and
- when the initial beach profile is steeper.

Since the higher parts of the beach are usually steeper than the lower parts, this in turn means, that the erosion is made more severe if the big waves coincide with a very high tide or a storm surge.

The time scale of profile adjustment is not universally defined. That is, some aspects of profile adjustment will quickly reach their asymptotic value while others take much longer. The total sand volume eroded from above the SWL in a storm approaches its asymptotic value on a time scale of only 2–3 hours according to the small ($H \approx 0.1\text{m}$) scale experiments by Vellinga (1986). Thus, the damage can be done during the peak of a typical storm, see Figure 1.6.7.

289

The shape of the deeper part of the profile takes much longer to adjust. The small scale experiments of Swart (1974) indicate that such adjustment occurs on a time scale more like 40 to 200 hours.

7.2.6 Bars

Bars on beaches occur with a large range of lengths and steepnesses. They are formed when the sediment transport converges as it does under the breakpoint in Figure 7.2.6. According to Dean and Dalrymple (2002) this type of bars are usually present if $\dfrac{gH_o^2}{w_s^3 T} > 10400$.

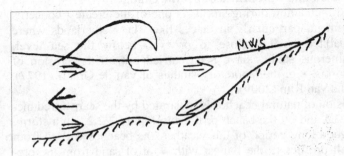

Figure 7.2.6: Breakpoint bars are generated by convergence of near-bed sediment transport under the breakpoint.

A similar convergence can be generated with wave reflection, instead of wave breaking, as shown by the small scale experiment in Figure 7.2.7.

The convection cells generated by wave reflection are periodic with length $L/2$. Partially reflected waves are often observed to generate such bar systems in the field.

The flat dissipative beach profile from Goolwa in Figure 7.2.2 shows multiple bars. On such a beach, large waves may break and reform several times on the way to the shoreline. The positions and spacings of such bars have been connected with standing edge waves (Short 1975) as well as with break-point positions (Dean et al. 1992).

The "intermediate" beaches: Mid Seven Mile and Eastern Beach in Figure 7.2.2 show a single bar, which moves considerably. Typically the bar will move rapidly offshore to the outer breakpoint of the large waves during a storm and then return slowly towards the shore during fair weather. If the

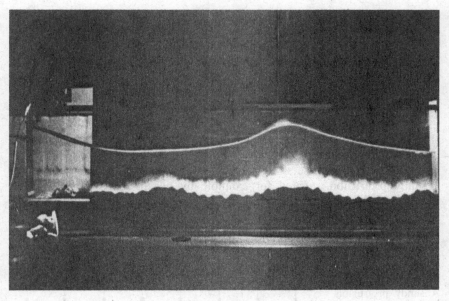

Figure 7.2.7: Standing waves drive circulation cells of length $L/2$ which generate periodic bars. Nielsen (1979). $T \approx 0.5$s, $h \approx 0.15$m, $d_{50} = 0.082$mm.

small waves persist and are small enough the bar may weld on the beach and become part of the berm.

The speed of landward bar migration V_{bar} was studied by Sunamura and Takeda (1984), who recommended the following formula

$$V_{bar} = 2 \times 10^{-11} \frac{w_o d_{50}}{\text{bar relief}} \left(\frac{H_s}{d_{50}} \right)^3 \qquad (7.2.4)$$

Flat macro-tidal beaches like Nudgee Beach Brisbane, Figure 7.2.5, has several offshore bars which are probably generated by other processes than wave breaking.

Nordstrom and Roman (1996) describe estuarine shores, which include macro-tidal low energy beaches, in great detail.

7.3   BEACH PLAN-FORMS

In a long term equilibrium situation, the shoreline will everywhere align itself so that the breaker line is parallel to the shore. It is a consequence of the fact that the longshore sediment transport rate is proportional to the angle between bottom contours and the breaker crest, cf Section 6.10, that the coastal plan-shape will change towards one where the breaker crests are everywhere parallel with the shoreline.

A special geometrical framework for constructing the corresponding shorelines, near dominant headlands, so-called ζ-shaped or crenulated beaches, was presented by Hsu et al. (1998).

For an otherwise straight coast with the waves coming straight in, a sandy protrusion, created for example by beach nourishment, will be flattened by the longshore sediment transport over time. See Figure 7.3.1.

Figure 7.3.1:
A sandy protrusion on a straight equilibrium shoreline, perhaps generated by localised beach nourishment, will be flattened by the longshore sediment transport over time.

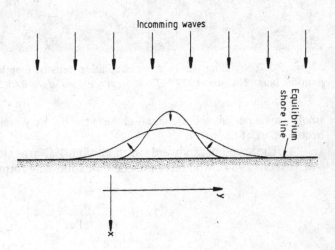

This flattening process (the gradual loss of the benefit of the nourishment) can be modelled by combining a formula for longshore sediment transport Q_y [e g, the CERC formula (6.10.1)] with the continuity principle. This leads to the following diffusion equation for the shoreline coordinate x_s

$$\frac{\partial x_s}{\partial t} = D \frac{\partial^2 x_s}{\partial y^2} \qquad (7.3.1)$$

see Pelnard-Considere (1956) or Dean (1998). The shoreline diffusivity is given by

$$D = \frac{KH_b^{5/2}\sqrt{\dfrac{g}{\gamma_b}}}{8(s-1)(1-p)(h_{cl}+B)} \tag{7.3.2}$$

where s is the specific gravity of the sand, p is the porosity, h_{cl} is the depth of closure and B is the berm height. For details of the constant K, see Section 6.10.

Equation (7.3.1) indicates that the morphological time scale T_{morf} for flattening of the protrusion in Figure 7.3.2, and thus, for the useful lifetime of the nourishment is

$$T_{morf} = \frac{L_n^2}{D} \propto \frac{L^2\left(h_{cl}+B\right)}{H_b^{5/2}\sqrt{g}} \tag{7.3.3}$$

where L_n is the length scale of the nourishment. It will be understood from the following derivation of the underlying continuity equation that equations (7.3.1) to (7.3.3) are based on the assumption that the beach profile shape is fixed.

7.4   SEDIMENT BUDGETS

The sediment budget is an important tool in coastal engineering in relation to sand bypassing and beach nourishment projects. It ties together the rates of shoreline retreat with gradients in the longshore sediment transport, sediment gain/loss from tidal inlets and sediment supply from rivers.

7.4.1   The 2D vertical sediment budget

In the simple situation where only shorenormal sediment transport is considered the book keeping is simple. If the onshore transport rate is Q_x [m²/s] (solid volume) and the active profile is between a depth of closure h_{cl} and a berm height B where it keeps a constant shape, the shoreline retreat $\dfrac{\partial x_s}{\partial t}$ will be given by

$$(1-p)\left(h_{cl}+B\right)\frac{\partial x_s}{\partial t} = -Q_x \tag{7.4.1}$$

293

The porosity p is usually in the range $p \in [0.3; 0.4]$, usually greater in wind blown sand than in sand that has been deposited by waves.

7.4.2 Bruun's rule for shoreline retreat due to sea-level rise

Bruun (1962) realised that if the sea-level rises by an amount Δ_z and if the sand, required for establishing the new, higher equilibrium profile, can only come from a shoreline retreat Δ_x, then, the shoreline retreat is given by

$$\Delta_x = \frac{\text{width of the active profile}}{\text{height of the active profile}} \Delta_z = \frac{x(B) - x(h_{cl})}{h_{cl} + B} \Delta_z \qquad (7.4.2)$$

This is obtained by equating the two rectangle areas $[h_{cl} + B]\Delta_x$ and $[x(B) - x(h_{cl})]\Delta_z$ in Figure 7.4.1. That is, the shoreline retreat equals the sea-level rise divided by the average slope of the active profile.

$$\Delta_x = \frac{\Delta_z}{\text{height of the active profile} / \text{width of the active profile}} \qquad (7.4.3)$$

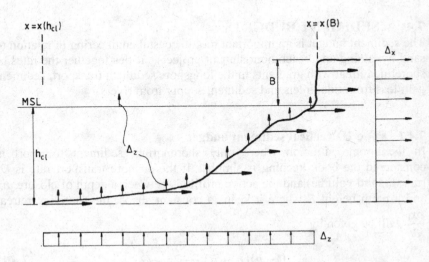

Figure 7.4.1: Bruun's (1962) rule for shoreline retreat due to sea-level rise.

7.4.3 3D sediment budgets

In analogy with the 2D situation in Figure 7.4.1, it is easily realised that the continuity equation for the general 3D situation in Figure 7.4.2 is

$$(1-p)(h_{cl}+B)\frac{\partial x_s}{\partial t} = -Q_x + \frac{\partial Q_y}{\partial y} + Q_{sinks} - Q_{sources} \qquad (7.4.4)$$

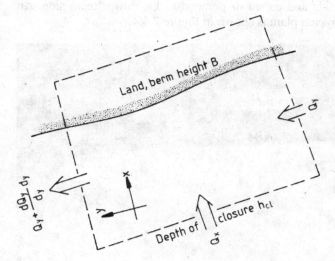

Figure 7.4.2: Control volume for the sediment budget and shoreline retreat equation (7.4.4). River mouths, marinas and/or lagoon entrances along the coast may act as sediment sinks or sources.

Figure 7.4.3: Lee side erosion downdrift of the groyne at the Mar Chiquita Inlet, Argentina. The littoral drift is from the top left towards the bottom right.

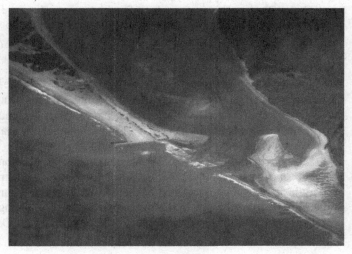

7.4.4   Lee side erosion

If a littoral drift barrier is erected, e g, a groyne as in Figure 7.4.3 or, a pair of training walls at a river entrance the sand will build up on the updrift side while the downdrift side will suffer erosion. These shoreline changes can be estimated using Equation (7.4.4) with $Q_y = 0$ at the barrier.

The adverse effects of necessary littoral barriers are nowadays addressed by sand bypassing. That is, the littoral drift material is picked up from the upstream side and carted or pumped to the downstream side. An example of a fixed bypass plant is shown in Figure 7.4.4.

Figure 7.4.4: The sand bypass system at the Tweed River on the Queensland - New South Wales Border. This system of ten fixed jet pumps along the pier is designed to bypass a natural littoral drift of the order 500,000m^3 per year.

7.4.5   Beach realignment due to climate change

The fact that shoreline retreat is likely in many areas, if the sea-level rises, has been well understood since the launching of the Bruun rule (Bruun 1962). Less appreciated is the fact that many sandy beaches will undergo significant realignments if the wind climate changes so that the dominant storm waves change direction as in Figure 7.4.5.

This realignment may involve many metres of shoreline recession near one end of long embayments. Also, tombolos and istmusses may be broken through as a consequence of the dominant waves changing direction. Some indication of the pending wind and wave climate changes can perhaps be gained from studying correlations between the Southern Oscillation Index (SOI) and storm severity as done by You and Lord (2005).

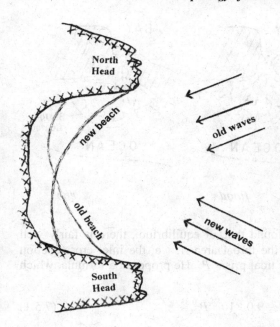

Figure 7.4.5:
Beach realignment due to change of wave direction. People at the northern end of the embayment have benefited from wave/beach realignment while those at the southern end have suffered serious erosion.

7.5   MORPHOLOGY OF TIDAL INLETS

Tidal inlets are complex dynamic environments in their natural state, see for example Figure 1.10.1 and the details of the processes forming the flood tide delta inside the entrance are far beyond the current state of the art of computer modelling.

Many tidal inlets on sandy coasts with littoral drift will migrate in the drift direction between major floods and then break through the spit to take a shorter flow path during floods. The banks and channels of inlets like Figure 1.10.1 are difficult to navigate not least because they can change quite rapidly during energetic flow conditions. Hence many river entrances have been extensively regulated to improve navigability and protect properties against erosion.

Some of the earliest quantitative work on entrance morphology was that of O'Brien (1931, 1969) and of Bruun (1968).

It is perhaps surprising that some natural sandy inlets are as narrow as they are. The mechanism for keeping the opening narrow was illustrated by O'Brien, see Figure 7.5.1.

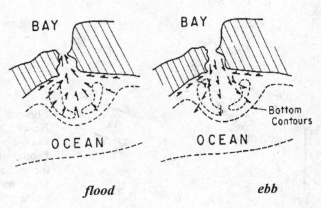

Figure 7.5.1:
The fact that return flow eddies form during ebb means that the nearshore flow is always towards the inlet. This keeps untrained inlets fairly narrow. After O'Brien (1969).

O'Brien (1969) also found that, at equilibrium, there is fairly well defined relationship between the throat area A, i e, the inlet cross section below mean sea level, and the tidal prism P. He proposed a formula, which in SI-units is

$$A = 9.0 \times 10^{-4} P^{0.85} \qquad (7.5.1)$$

7.6 DELTAS

Where the freshwater outflow is dominant over the tidal flows and where waves are too small to significantly rework the sediment, a river may build a delta out into the sea (or lake). Famous examples are the Nile, the Mississippi the Ganges-Brahmaputra and Papua New Guinea's Fly River.

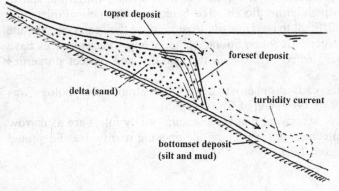

Figure 7.6.1:
Delta sedimentation. After Kostic and Parker (2003). The slopes are strongly exaggerated.

The sedimentation process has been logically schematised by Kostic and Parker (2003) as in Figure 7.6.1.

The gradual slowing of the river flow, as the slope gets flatter, leads to the deposition of the coarser sediment in the *topset*. Sedimentation strata in the *topset* will slope only mildly like the river bottom.

During steady or falling sea level, the river will usually still be transporting some sand or silt at the *plunge point*. This material will then avalanche down. This material is deposited in the *foreset* and the *plunge point* gradually moves seaward. The *foreset* is characterised by steeply inclined sediment strata.

From the base of the *foreset*, mud and fine silt may continue seaward carried by *turbity currents* until these finally lose their momentum at the abyssal floor. Along the way, the fine sediment is gradually deposited in the *bottomset*. The strata slope in the *bottomset* is much smaller than in the *foreset*.

7.7 CORAL CAYS AND ATOLLS

Coral cays and atolls owe their existence to living creatures, namely corals. The corals synthesise calcium carbonate, which initially forms the coral reefs and in its recycled form as carbonate sand or gravel may form coral cays.

Coral reefs are common in tropical and subtropical seas with low turbidity and low nutrient content. The corals cannot survive drying and the height of living corals is therefore restricted to a few tens of centimetres above low tide, where wave splashing can keep them permanently moist.

If the tidal range is small, e g <0.5m, as in the middle of the Pacific Ocean, the wave splashing may be sufficient for the living corals to extend above the high tide level. The lagoon is then more or less closed throughout the tidal cycle, and the flushing of the lagoon must be driven by the waves as described in Section 4.8.

Where the tidal range is larger, e g three to five metres, as in parts of the Australian Great Barrier Reef, the high tide will be metres above the living coral. In this scenario the flushing is provided by the tide and wind driven currents. Tidal flushing may also be dominant for ocean atolls with a less complete rim than Manihiki, e g, Penrhyn in the Cook Islands.

Corals need light and will therefore only grow in moderate depths usually along the edge of continents or as fringing reefs around islands, Tahiti and Rarotonga are surrounded by almost unbroken fringing reefs. Atolls have usually started as fringing reefs within which the volcanic island has been eroded away or become submerged due to cooling of the substrates. In the case of submergence due to cooling, the reef will usually keep up with the relative sea-level rise and remain a few tens of centimetres above low tide. The interior of the lagoon may then continually be filled with coral sand if waves and tidal currents are able to transport the coral sand across the reef flat and into the lagoon. If the tidal range is small, this may not happen. Instead the coral gravel and sand will form beaches which may become vegetated by coconut palms. Behind such a more or less continuous barrier, the atoll lagoon may become very deep as subsidence continues.

The Cook Islands show a sequence of young volcanic islands with fringing reefs in the south, e g Rarotonga, and very mature atolls in the north, e g, Manihiki. The top of the volcanic substrate of Manihiki is covered by 500m of limestone and, on top of that, the lagoon is up to 70m deep.

Coral cays may form where the production of coral sand and gravel is more than sufficient for filling the lagoon. Australia's Great Barrier Reef has many coral cays which have become prominent tourist destinations, e g, Heron Island and Green Island. Much of the carbonate sand in the cays gets cemented into beach rock and thus stabilised. However, in the uncemented state the cays are very vulnerable to tropical cyclones.

Figure 7.7.1: Green turtles on Raine Island of the NE coast of Australia, November 2006. In recent years the nesting success of these turtles has been reduced due to many eggs drowning, i e, not enough unsaturated sand above the watertable. Photos courtesy of Paul Guard.

8 Coastal groundwater dynamics

8.1 INTRODUCTION

The groundwater dynamics and salinity in aquifers, which border an ocean or an estuary, will be affected by the tides and the waves in ways that are important for water supply and water quality.

If the aquifer borders the beach face, as in Figure 8.1.1, the processes of wave runup and setup combined with the asymmetry of the tidal in- and out-flow through the slope generate a quasi-steady watertable overheight η^+ above the MSL, which is important in formulating the ocean boundary conditions for regional groundwater models.

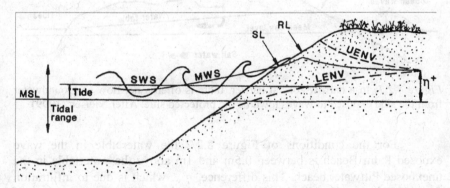

Figure 8.1.1: Wave runup and nearshore wave setup will elevate the time-mean watertable in aquifers which border the beach face. Depending on beach topography, tide range and wave conditions, η^+ may be more than a metre, which is clearly significant for water and salinity management in aquifers near the coast. Close to the beach, the watertable will oscillate between its upper and lower envelopes, UENV and LENV. See also Figure 2.3.1 for swash zone terminology. A complete expression for η^+ including non-linear flow effects is given by (8.4.15).

8.2   QUASI-STEADY EFFECTS OF WAVES AND TIDES ON THE INLAND WATERTABLE

In a narrow coastal barrier, where the watertable overheight due to rainfall is small compared with the wave runup height, the watertable may slope landward monotonically and a net flow of groundwater towards the continent exists. See Figure 8.2.1.

The scenario of Figure 8.2.1 generates important environmental concerns:

- Oil and other chemicals washed up on the beach face must be mechanically removed rather than dissolved as dissolved chemicals will enter the aquifer with the runup infiltration, and once inside the aquifer, they will be very difficult to remove.
- The flow of waste water towards the protected side must be kept in mind in relation to the siting of oyster farms and the like.

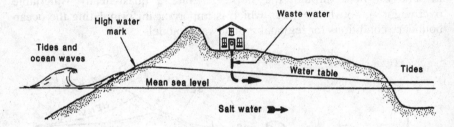

Figure 8.2.1: In a narrow coastal barrier, there is often a net flow of groundwater from the side exposed to waves towards the protected side. After Nielsen (1999).

For the conditions of Figure 8.2.2 the watertable in the wave exposed Palm Beach is between 0.5m and 1m above the watertable in the unexposed Pittwater beach. This difference, η^+_w, which is due to infiltration from wave runup is very significant in connection with the management of water supply and salinity in coastal aquifers.

Since η^+_w is generated by infiltration from wave runup it can be expected to scale on the vertical runup scale, cf Section 2.4, and Nielsen (1999) recommended

$$\eta^+_w \approx 0.44\sqrt{H_{o,rms}L_o}\tan\beta \qquad (8.2.1)$$

based on laboratory and field data available at the time.

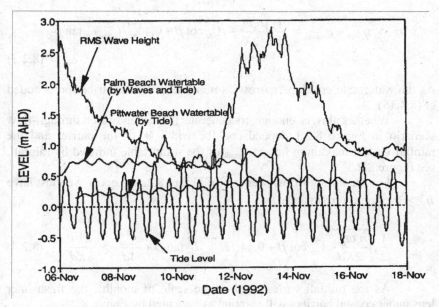

Figure 8.2.2: Watertable measurements taken just above the high water mark on the two sides of the Palm Beach Isthmus, north of Sydney, Australia. Palm Beach is exposed to the ocean waves while Pittwater Beach is not. Both sides are exposed to the same tide.

While the ocean tide in Figure 8.2.2 has time mean zero, the Pittwater Beach watertable is clearly on the average above MSL. This average overheight η^+_{slope} is, in this case, not due to rainfall but to the slope of the sand surface.

For a simple harmonic tide with amplitude A_{tide} and angular frequency ω_{tide} in a shallow aquifer with depth d below MSL, hydraulic conductivity K and effective porosity n_e, Nielsen (1990) derived

$$\eta^+_{\text{slope}} = \frac{1}{2}kA^2_{\text{tide}}\cot\beta = \frac{1}{2}\sqrt{\frac{n_e\omega_{\text{tide}}}{2Kd}}A^2_{\text{tide}}\cot\beta \qquad (8.2.2)$$

This is an accurate estimate for steep beaches but an overestimate for flat beaches since $\cot\beta\to\infty$ for $\beta\to0°$, while η^+_{slope} should clearly not become greater than A_{tide}.

In summary, the time-mean watertable close to a beach with waves and tide will be elevated above MSL by

$$\eta^+ = \eta^+_{\text{slope}} + \eta^+_{\text{wave}} \approx \frac{1}{2}\sqrt{\frac{n_e\omega_{\text{tide}}}{2Kd}}\, A^2_{\text{tide}}\cot\beta + 0.44\sqrt{H_{\text{o,rms}}L_o}\,\tan\beta$$

$$(8.2.3)$$

As the watertable envelope narrows a non-linear flow contribution is added cf (8.4.15).

Whether this is enough to generate the groundwater through-flow scenario in Figure 8.2.1 depends on the width W of the barrier and the rainfall via the maximum height η_{peak} of the water table induced by rainfall, see Figure 8.2.3.

For the through-flow scenario of Figure 8.2.1 to occur we must have $\eta^+ \gtrsim \eta_{\text{peak}}$ or approximately

$$\frac{1}{2}\sqrt{\frac{n_e\omega_{\text{tide}}}{2Kd}}\, A^2_{\text{tide}}\cot\beta + 0.44\sqrt{H_{\text{o,rms}}L_o}\,\tan\beta + \frac{A^2_{\text{tide}}}{4d} \gtrsim \frac{W^2 i}{8Kd} \qquad (8.2.4)$$

As the rainfall varies on the time scale of months, the freshwater lens under coastal barriers will respond as indicated by Figure 8.2.4.

For narrow barriers like the one in Figure 8.2.4, the thickness of the freshwater lens generally tapers to zero near the ocean beach, but its shape changes in response to major rainfall and wave events. After a few months of dry weather the freshwater lens is reduced and pushed towards the "protected side" by the wave induced groundwater through-flow. The thickness of the freshwater lens is calculated as

Figure 8.2.3:
The maximum elevation of the watertable in a barrier of width W is approximately

$$\eta_{\text{peak}} = d\left(\sqrt{1+\frac{W^2 i}{4Kd^2}}-1\right)$$

$$\approx \frac{W^2 i}{8Kd}.$$

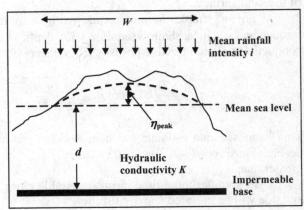

304

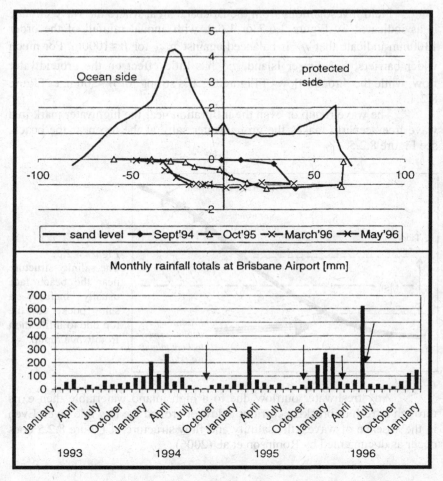

Figure 8.2.4: Rainfall time series (bottom) and corresponding freshwater lenses δ_{fresh} under Bribie Island near Brisbane. Elevations and lens thickness in metres.

$$\delta_{\text{fresh}} = \int \frac{c_{\text{sea}} - c}{c_{\text{sea}}} \, dz \tag{8.2.5}$$

from the salinity profile $c(z)$, which generally does not display a sharp interface in these scenarios. Data from Nielsen (1999).

305

Field investigations from the Brisbane area, where the wave climate is as indicated by Figure 1.6.7 or 1.6.9, with annual rainfall of the order 1300mm indicate that η^+ is balanced against η_{peak} for $W \approx 1000$m. For much wider barriers, like Fraser Island, η^+ has little effect on the groundwater flow, while the through-flow of Figure 8.2.1 is strong for $W < 200$m, cf Figure 8.2.4.

The wave runup or even the infiltration near the highwater mark in a wave-free scenario makes the groundwater salty at the top near the beach, see Figure 8.2.5.

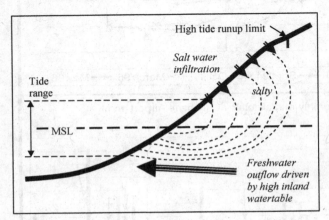

Figure 8.2.5: The salinity structure near the beach face usually includes a salty pocket at the top due to infiltration from wave runup.

Any freshwater outflow due to a high inland watertable then exits through a fresh or brackish corridor below or around the low tide level. Even in the absence of waves, the salinity and flow structure of Figure 8.2.5 does occur as documented by Robinson et al. (2006).

8.3 WATERTABLE VARIATION AND WATER EXCHANGE AT THE BEACH FACE

The pore water pressures in the sand just landward of the runup limit exhibit vigorous oscillations which are related to individual runup events. To some extent these pressure fluctuations are synonymous with watertable fluctuations. Not quite however, since the oscillating pressures are not necessarily hydrostatic.

The relations between watertable changes and flow velocities are also far from trivial where infiltration is involved, see for example Figure 8.3.1.

Rain

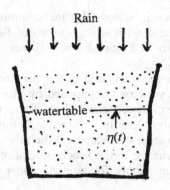

Figure 8.3.1:
Rainfall fills the pores in the sand and makes the watertable rise. At the same time however, all water velocities in the sand are zero or downwards, i e,

$$\frac{\partial \eta}{\partial t} > 0 \quad \text{while} \quad w \leq 0.$$

The watertable dynamics under the beach are indicated by the movements of the border *GD* between the lower *G*lassy area and the upper *D*ry looking area on the beach face, Figure 8.3.2. *GD* is the *exit line* where the watertable outcrops onto the beach face. *GD* sits higher in the runup distribution and may even be above the runup limit during falling tides. Turner (1993) gives some detailed observations from macro-tidal beaches.

Figure 8.3.2: Seepage face during falling tide on a fairly flat, $\tan \beta_F \approx 0.06$, fine grained beach, $d_{50} \approx 0.2$mm. The border between the glassy (wet) and dry looking areas is the exit line where the watertable outcrops onto the beach.

While fast movements of the beach watertable are commonly observed, these are not necessarily associated with strong in or out flow. Meniscus formation plays a role as shown in Figure 8.3.4. Turner and Nielsen (1997) give experimental details.

The watertable fluctuations which are generated under the swash zone travel landwards as watertable waves which are dampened by viscous dissipation, most strongly for the higher frequencies. High frequency watertable fluctuations are thus not significantly influencing inland groundwater dynamics. It is however of interest to understand the relations between porewater pressures and water exchange in the beach in relation to swash zone sediment transport and oxygen supply for animals under the swash.

8.3.1 Pressure fluctuations under the beach face

Consider the scenario of Figure 8.3.3 where the swash depth and the porewater pressure in the sand below are monitored.

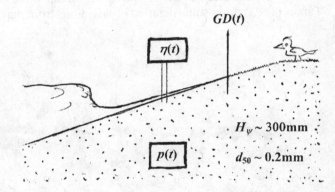

Figure 8.3.3: Simultaneous measurements of swash depth η and porewater pressure p.

The pressure head, $p/\rho g$, fluctuations, which indicate the watertable fluctuations may be up to the order of the capillary rise $H_\psi \sim$ tens of centimetres for typical beach sand, while the surface elevation η only reaches a few centimetres above the sand. This is because the watertable can be a full capillary height H_ψ below the sand surface due to miniscus suction at the top of a saturated sand body, see Figure 8.3.4. Cartwright et al. (2006) present field measurements.

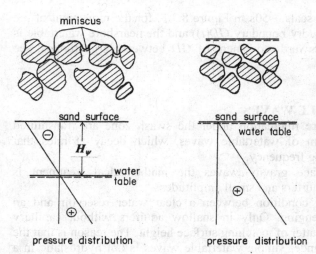

Figure 8.3.4:
With fully formed miniscusses, the watertable is $H_\psi \sim$ tens of centimetres below the sand surface, while it only takes half a grain diameter ~ 0.1mm of water to eliminate the miniscusses and bring the watertable to the sand surface. This corresponds to the effective porosity defined by (8.4.9) being as tiny as $\sim d_{50}/H_\psi$.

The dominance of the meniscus formation/disappearance as a driver of the watertable and associated pore water pressure fluctuations near the beach is documented by the spectra in Figure 8.3.5.

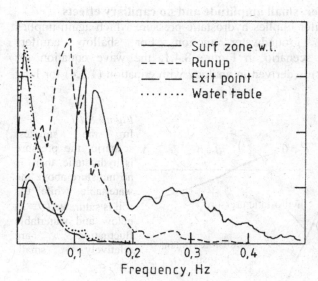

Figure 8.3.5:
The spectrum of watertable fluctuations a short distance landward of the runup limit is very similar to that of the movement of the exit point or the "glassy/dry boundary" *GD* in Figure 8.3.2. Both the surf zone waterlevel and the runup, i e the waterline motion along the beachface have much more high frequency content. Data courtesy of Dr Ian Turner.

The long time scale, ~50s in Figure 8.3.5, for the movement of the exit point (= the glassy/dry boundary $GD(t)$) and the nearshore watertable is that of the gradual seaward movement of GD between large runups, see Figure 8.3.2.

8.4 WATERTABLE WAVES

The porewater pressure fluctuation under the swash zone are transmitted landwards in the form of watertable waves, which decay at rates that increase with increasing frequency.

As with surface gravity waves the mathematical treatment is simplest for shallow aquifers and small amplitudes.

The boundary condition between a clear water reservoir and an aquifer is very challenging. Only in shallow aquifers without capillary effects is it a simple matter of matching surface height. The reason is that the pressure distribution under simple watertable waves is not hydrostatic in a non-shallow aquifer. Hence the usually hydrostatic pressure in the reservoir cannot be matched by a single wave mode. Capillary effects at the interface give further complications.

8.4.1 Shallow aquifer, small amplitude and no capillary effects

The term shallow aquifer implies hydrostatic pressure which again implies essentialy horizontal groundwater motion. For shallow aquifers corresponding to the scenario in Figure 8.4.1, the wave equation for watertable waves $\eta(x,t)$ is derived in analogy with equation (1.3.4) for long surface gravity waves.

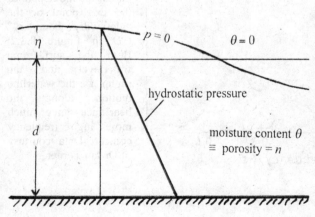

Figure 8.4.1: In the simplest scenario, the pressure is hydrostatic, there is no moisture above the watertable while the soil is saturated ($\theta \equiv n$) below and watertable fluctuations are relatively small, $|\eta| << d$.

The difference is that Newton II is replaced by Darcy's law

$$u = -K\frac{\partial \eta}{\partial x} \tag{8.4.1}$$

where u is the discharge per unit area and K is the hydraulic conductivity. The continuity equation is

$$n\frac{\partial h}{\partial t} = -\frac{\partial}{\partial x}[u(d+\eta)] \approx -d\frac{\partial u}{\partial x} \tag{8.4.2}$$

The factor n, the porosity on the left expresses that only the pore space is filled with water. Here u can be eliminated to give

$$\frac{\partial \eta}{\partial t} = \frac{Kd}{n}\frac{\partial^2 \eta}{\partial x^2} \tag{8.4.3}$$

which has decaying solutions of the form

$$\eta(x,t) = A_o e^{-kx} \cos(\omega t - kx)$$

$$= \text{Re}\{A_o e^{-(1+i)kx} e^{i\omega t}\} \tag{8.4.4}$$

where the wave number k is given by the shallow aquifer dispersion relation

$$k = \sqrt{\frac{n\omega}{2Kd}} \tag{8.4.5}$$

in which ω is the angular frequency, K is the hydraulic conductivity and d is the undisturbed aquifer depth.

The solution (8.4.4) can be matched with a tidal reservoir with hydrostatic pressure and surface elevation $A_o\cos\omega t$ at $x=0$.

8.4.2 Capillary effects, partial saturation and effective porosity

The watertable dynamics are somewhat more complicated in real aquifers where the moisture content is not zero above the watertable but may be distributed qualitatively as in Figure 8.4.2. The total drainable moisture content is $(\theta_s-\theta_r)h_{tot}$ corresponding to

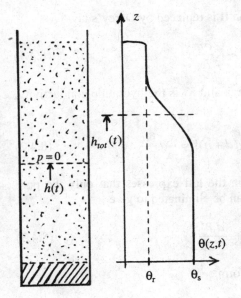

Figure 8.4.2:
Below the watertable $z = h$, the moisture content is uniform at its saturated value $\theta_s \equiv n$. Far above the watertable, θ has a residual value θ_r corresponding to the porewater which will not drain away under gravity. In between, $\theta = \theta(z,t)$ varies in a way which depends on the history of $h(t)$.

$$h_{\text{tot}} = \int \frac{\theta - \theta_r}{\theta_s - \theta_r} dz \qquad (8.4.6)$$

Allowing for real moisture distributions and for $h \neq h_{\text{tot}}$ in general requires modification of the continuity equation (8.4.2) to

$$(\theta_s - \theta_r)\frac{\partial h_{\text{tot}}}{\partial t} \approx -d\frac{\partial u}{\partial x} \qquad (8.4.7)$$

while Darcy's law in the form (8.4.1) still holds as long as the aquifer is shallow $\sim n\omega d/K \ll 1$. Eliminating u between the two equations now gives

$$(\theta_s - \theta_r)\frac{\partial h_{\text{tot}}}{\partial t} = Kd\frac{\partial^2 \eta}{\partial x^2} \qquad (8.4.8)$$

where we need to eliminate one of the dependent variables η or h_{tot}. This can be done by introducing the effective porosity n_e defined by

$$n_e \frac{\partial \eta}{\partial t} = (\theta_s - \theta_r)\frac{\partial h_{tot}}{\partial t} \qquad (8.4.9)$$

or $\dfrac{n_e}{\theta_s - \theta_r} = \dfrac{\partial h_{tot}}{\partial \eta}$, which by insertion into (8.4.8) gives

$$n_e \frac{\partial \eta}{\partial t} = Kd \frac{\partial^2 \eta}{\partial x^2} \qquad (8.4.10)$$

In analogy with the simplified capillarity free scenario considered above, this has solutions of the form

$$\eta(x,t) = A_o e^{-k_r x} \cos(\omega t - k_i x)$$

$$= \mathrm{Re}\{A_o e^{-kx} e^{i\omega t}\} \qquad (8.4.4)$$

where the complex wave number k is given by

$$k = k_r + i k_i = \sqrt{i \frac{n_e \omega}{Kd}} \qquad (8.4.11)$$

Cartwright et al. (2005) investigated the behaviour of n_e in relation to simple harmonic watertable oscillations in several sand-like materials and recommended

$$n_e = \frac{n}{1 + 2.5\left(i\dfrac{n\omega H_\psi}{K}\right)^{2/3}} \qquad (8.4.12)$$

The fact that n_e turns out to be complex corresponds to the fact that h_{tot} and η are generally not in phase under an oscillating watertable. The negative argument of n_e as given by (8.4.12) means that the total moisture is generally delayed relative to the watertable.

8.4.3 Shallow aquifer, finite amplitude

Finite amplitude, shallow aquifer solutions are not available in analytical form, but may easily be obtained by numerical integration of the so-called Boussinesq equation

$$n\frac{\partial \eta}{\partial t} = K\frac{\partial^2}{\partial x^2}\left(\frac{(d+\eta)^2}{2}\right) \tag{8.4.13}$$

which results from (8.4.1) and (8.4.2) without applying the small amplitude assumption.

The non-linearity of this equation gives an extra overheight $\bar{\eta}_{nl}$ of the inland watertable compared with MSL. Philip (1973) showed by time averaging of (8.4.13) that, with the seaward boundary condition $\eta(0,t) = A\cos\omega t$, the contribution from non-linearity to the asymptotic inland overheight of the watertable is

$$\bar{\eta}_{nl} = \sqrt{d^2 + \frac{A^2}{2}} - d \approx \frac{A^2}{4d} \tag{8.4.14}$$

This is to be added to the slope- and wave runup effects discussed in Section 8.1. That is, far inland, we expect the steady watertable overheight

$$\eta^+ = \eta^+_{\text{slope}} + \eta^+_{\text{wave}} + \bar{\eta}_{nl}$$

$$\approx \frac{1}{2}\sqrt{\frac{n\omega_{\text{tide}}}{2Kd}}\,A^2_{\text{tide}}\cot\beta + 0.44\sqrt{H_{o,\text{rms}}L_o}\tan\beta + \frac{A^2_{\text{tide}}}{4d} \tag{8.4.15}$$

in the absence of recharge and with negligible capillary effects.

The experimentally based complex n_e given by (8.4.12) is not easily incorporated in (8.4.13) to account for capillary effects. However, the simplified capillary fringe dynamics corresponding to the model of Green and Ampt (1913) where the sand is saturated up to the top of a capillary fringe where the pressure at the top is fixed at $-\rho g H_\psi$, can be included. The Green and Ampt model corresponds to

$$n_e = \frac{n}{1 + i\dfrac{\omega H_\psi}{K}} \tag{8.4.16}$$

cf Cartwright et al. (2005). We note that this form is similar, yet quite different from the experimentally based (8.4.12). By inserting (8.4.16) into (8.4.13) one gets

$$n\frac{\partial\eta}{\partial t} = K\left[1+i\frac{\omega H_\psi}{K}\right]\frac{\partial^2}{\partial x^2}\left(\frac{(d+\eta)^2}{2}\right) \qquad (8.4.17)$$

Here we can note that for simple harmonic watertable fluctuations $\sim e^{i\omega t}$ the factor $i\omega$ corresponds to $\dfrac{\partial}{\partial t}$ and hence (8.4.17) corresponds to

$$n\frac{\partial\eta}{\partial t} = K\left[1+\frac{H_\psi}{K}\frac{\partial}{\partial t}\right]\frac{\partial^2}{\partial x^2}\left(\frac{(d+\eta)^2}{2}\right) \qquad (8.4.18)$$

a form which was derived along different lines by Barry et al. (1996).

8.4.4 Finite aquifer depth, small amplitude

If the aquifer is not shallow, the pressure field under a watertable wave is not hydrostatic. There are however still simple harmonic small amplitude solutions of a form

$$\eta(x,t) = Ae^{-k_r x}\cos(\omega t - k_i x) \qquad (8.4.19)$$

for the watertable. This is similar to the shallow aquifer solution (8.4.4). However, the real and imaginary parts of the complex wave number $k = k_r + ik_i$ are no longer equal, not even in the absense of capillary effects. There are infinitely many wave modes for each frequency. The modes have different wave numbers, which all satisfy the dispersion relation

$$kd\,\tan kd = i\frac{n_e\omega d}{K} \qquad (8.4.20)$$

cf Nielsen et al. (1997). This dispersion relation accounts for capillary effects if the expression (8.4.12) is used for n_e. For each mode, the corresponding oscillatory pressure $\tilde{p}$ varies in amplitude and phase through a vertical as

$$\frac{\tilde{p}(x,z,t)}{\rho g\eta(x,t)} = \frac{\cos kz}{\cos kd} \quad \text{where } z=0 \text{ at the base.} \qquad (8.4.21)$$

This is somewhat analogous with the pressure variation under sine waves, Equation (1.4.33). However, the swapping of cos for cosh and having complex wave numbers means that under the groundwater waves, the pressure fluctuations are strongest at the bottom and they happen earlier at

315

greater depth. Physically, the phase lead and greater pressure amplitude at the bed is due to the fact that these waves are decaying in the x-direction and that events at depth become dominated by the higher amplitude activities on the left (for waves propagating from left to right).

This non-hydrostatic behaviour of watertable wave modes at non-shallow depths means that no single mode can match the hydrostatic pressure in a bordering clear water reservoir. The boundary condition must be matched by a suitable combination of modes.

For the small amplitude case, Nielsen et al. (1997) devised an infinite series solution by which the hydrostatic boundary condition

$$\tilde{p}(0,z,t) = \begin{cases} \rho g A \cos \omega t & \text{for } 0 < z < d \\ 0 & \text{for } \quad z \ge d \end{cases} \qquad (8.4.22)$$

is satisfied. This solution is based on the fact that (8.4.20)+(8.4.21) represent a set of orthogonal functions. It is however only valid for small amplitude and a vertical interface. The watertable dynamics near a real, sloping interface, which is also influenced by capillary suction above the seepage face is beyond analytical description. A detailed experimental investigation was presented by Cartwright et al. (2004). Their data indicate the qualitative scenario of Figure 8.4.3.

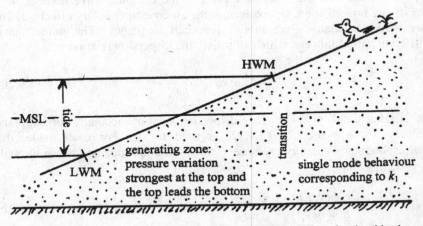

Figure 8.4.3: The pressure fluctuations under a real beach driven by the tide show a *generating zone* between the HWM and the LWM where fluctuations are strongest at the top. Then a *transition* zone landward of which the groundwater waves asymptote towards a single mode of the form (8.4.16) corresponding to the lowest mode, i e, the one with the smallest k_r and hence the slowest decay.

316

References

Aagaard, T, B Greenwood and J Nielsen (2000): Bed level changes and megaripple migration on a barred beach. *J Coastal Res, ICS 2000 Proceedings*, pp 110-116.

Allen, J R L (1984): *Sedimentary Structures: Their Character and Physical Basis.* Elsevier, 663pp.

Alonso, M and E J Finn (1968): *Fundamental University Physics, Vol 1 Mechanics*, Addison Wesley, 463pp.

Alsina, J M and T E Baldock (2007): Improved representation of breaking wave energy dissipation in parametric wave transformation models. *Coastal Engineering, Vol 54, No 10*, pp 765-769.

Amos, C L, A J Bowen, D A Huntley and C F M Lewis (1988): Ripple generation under the combined influence of waves and currents on the Canadian continental shelf. *Cont Shelf Res, Vol 8, No 10*, pp 1129-1153.

Arnott, R W and J B Southard (1990): Exploratory duct flow experiments on combined flow bed configurations and some implications for interpreting storm event stratification. *J Sedimentary Petrology, Vol 60 No 2*, pp 211-219.

Bagnold, R A (1956): The flow of cohesionless grains in fluids. *Phil Trans Roy Soc Lond, Vol 249*, pp 235-297.

Bagnold, R A (1962): Auto-suspension of transported sediment; turbidity currents. *Proc Roy Soc Lond, Vol A 265*, pp 315-319.

Bagnold, R A (1963): Mechanics of marine sedimentation. In N M Hill ed *The Sea, Vol 3*, Interscience N Y, pp 507-528.

Bailard, J A (1981): An energetics total load sediment transport model for a plane sloping beach. *J Geophys Res, Vol 86, No C11*, pp 10938-10954.

Bakker, W T and W G M van Kesteren (1986): The dynamics of oscillatory sheet flow. *Proc 20th Int Conf Coastal Eng*, Taipei, ASCE, pp 940-954.

Baldock T E (2006): Long wave generation by the shoaling and breaking of transient wave groups on a beach. *Proc Roy Soc Lond A, Vol 462*, pp 1853-1876.

Baldock, T E and P Holmes (1998): Seepage effects on sediment transport by waves and currents. *Proc 26th Int Conf Coastal Engineering*, Copenhagen, ASCE, pp 3601-3614.

Baldock, T E, M R Tomkins, P Nielsen and M G Hughes (2004): Settling velocity of sediments at high concentrations. *Coastal Engineering, Vol 51*, pp 91-100.

Barry, M, G N Ivey, K B Winters and J Imberger (2001): Measurements of diapycnal diffusivities in stratified fluids. *J Fluid Mech, Vol 442*, pp 267-291.

Bascom, W (1980): *Waves and Beaches*. Anchor Press/Doubleday, New York.

References

Barry, D A, S J Barry and J-Y Parlange (1996): Capillarity correction toperiodic solutions of the shallow flow approximation. In *Mixing in Estuaries and Coastal Seas, Coastal and Estuarine Studies*, edited by C B Pattiaratchi, A G U, Washington D C, pp 496-510,

Batchelor, G K (1967): *An Introduction to Fluid Dynamics*. Cambridge Univ. Press, 615pp.

Battjes, J A, H J Bakkenes, T T Janssen and A R van Dongeren (2004): Shoaling of subharmonic gravity waves. *J Geophysical Res, Vol 109, C02009*, 15pp.

Bayram, A, M Larson, H C Miller and N C Kraus (2001): Cross shore distribution of longshore sediment transport: Comparison between predictive formulas and field measurements. *Coastal Engineering, Vol 44, No 2*, pp 79-99.

Bear, J (1972): *Dynamics of Fluids in Porous Media*. Elsevier, New York, 764pp.

Benjamin, T B and M J Lighthill (1954): On cnoidal waves and bores. *Proc Roy Soc Lond A, Vol 224*, pp 448-460.

Berkhoff, J W C (1972): Computation of combined refraction-diffraction. *Proc 13th Int Conf Coastal Eng*, ASCE Reston VA, pp 471-490.

Bernard, E N, H O Mofjeld, V Titov, C E Synoloakis and F I Gonzales (2006): Tsunami: scientific frontiers, mitigation, forecasting and policy implications. *Phil Trans Roy Soc Lond, Vol A 364*, pp 1989-2007.

Bijker, E, W E van Hijum and P Vellinga (1976): Sand transport by waves. *Proc 15th ICCE*, Honolulu, pp 1149-1167.

Birkemeier, W A (1985): Field data on seaward limit of profile change. *J Waterways Port and Coastal Engineering*, ASCE, *Vol 111, No 3*, pp 598-602.

Black, K P (1994): Suspended sediment load during an asymmetrical wave cycle over a plane bed. *Coastal Engineering, Vol 23*, pp 95-114.

Bodge, K R (1992): Representing equilibrium beach profiles with an exponential expression. *J Coastal Res, Vol 8, No 1*, pp 47-55.

Bosman, J J and H Steetzel (1986): Time and bed averaged sediment concentrations under waves. *Proc 20th Int Conf Coastal Eng*, Taipei, ASCE, pp 986-1000.

Boussinesq, J (1872): Theorie des ondes et des remous qui se propagent le long d'uncanal rectangulaire horizontal, en communicant au liquide contenu dans ce canal des vitesses sensiblement pareilles de la surfaceau fond. *J de mathematique pures et appliquees, Deuzieme Serie, Vol 17*, pp 55-108.

Boussinesq, J (1877): *Essay sur la Theorie des Eaux courantes*. Paris.

Broccini, M and N Dodd (2008): Non-linear shallow water equation modelling for coastal engineering. *J Waterways Port and Coastal Engineering*, ASCE, *Vol 134, No 2*, pp 104-120.

Bruun, P (1968): *Stability of Tidal Inlets*. Elsevier, Developments in Geotechnical Engineering, Vol 23, 508pp.

References

Bryant, E (2008): *Tsumami: The Underrated Hazard*, 2nd edition. Springer, 330pp.

Buhr Hansen J and I A Svendsen (1979): Regular waves in shoaling water: Experimental data. Series Paper 21, ISVA, Tech Univ Denmark.

Buxbom, I P (2003): Large eddy simulation of the ventilated wave boundary layer. PhD Thesis, T U Denmark, 194pp.

Bye, J A T and A D Jenkins (2006): Drag coefficient reduction at high wind speeds. *J Geophysical Res, Vol 111, C03024*, doi:10.1029/2005JC0003114.

Byrne, R J (1969): Field occurrence of induced multiple gravity waves. *J Geophys Res, Vol 74, No 10*, pp 2590-2596.

Callaghan, D P, P Nielsen, N Cartwright, M R Gourlay and T E Baldock (2006): Atoll lagoon flushing forced by waves. *Coastal Engineering, Vol 53*, pp 691-704.

Callaghan, D P, J Callaghan, P Nielsen and T Baldock (2007): Generation of extreme wave conditions from an accelerating tropical cyclone. *Proc 30th Int Conf Coastal Eng, ASCE*, World Scientific, pp 752-760.

Carrier, G F and H P Greenspan (1958): Water waves of finite amplitude on a sloping beach. *J Fluid Mech, Vol 4*, pp 97-109.

Carslaw, H S and J C Jaeger (1959): *The Conduction of Heat in Solids*. Oxford University Press, London, 2nd ed., 510pp.

Carstens, M R, F M Neilson and H D Altinbilek (1969): Bedforms generated in the laboratory under an oscillatory flow. C E R C Tech Memo 28.

Cartwright N, L Li and P Nielsen (2004): Response of the salt-freshwater interface in a coastal aquifer to a wave-induced groundwater pulse: field observations and modelling, *Advances in Water Resources, Vol 27 No 3*, pp 297-303.

Cartwright, N, T E Baldock, P Nielsen D-S Jeng and L Tao (2006): Swash-aquifer interaction in the vicinity of the water table exit point on a sandy beach. *J Geophysical Res, Vol 111, C09035*, doi:10.1029/2005JC003149.

Cartwright N, Nielsen P and L Li (2004): Experimental observations of watertable waves in an unconfined aquifer with a sloping boundary, *Advances in Water Resources, Vol 27*, pp 991-1004.

Cartwright N, P Nielsen and P Perrochet (2005): Influence of capillarity on a simple harmonic oscillating watertable: Sand column experiments and modeling. *Water Resources Res, Vol 41, No 8 W08416*.

Cavaleri, L and S Zecchetto (1987): Reynolds stresses under wind waves. *J Geophysical Res, Vol 92, No C2*, pp 3894-3904.

Chanson, H (2001): An introduction to turbulent diffusion and dispersion in streams. Teaching notes, Department of Civil Engineering, The University of Queensland.

Chanson, H (2004): *Environmental Hydraulics of Open Channel Flows.* Elsevier Butterworth-Heinemann, Oxford, UK, 483pp.

Chanson, H (2004): *The Hydraulics of Open Channel Flows: An Introduction.* Butterworth-Heinemann, Oxford, UK, 2nd edition, 630pp.

319

References

Chanson, H, R Brown, J Feris and K Warburton (2003): A Hydraulic, Environmental and Ecological Assessment of a Sub-tropical Stream in Eastern Australia: Eprapah Creek, Victoria Point QLD on 4 April 2003. Report No. CH52/03, Dept. of Civil Engineering, The University of Queensland, Brisbane, Australia, June, 189pp.

Cherlet, J, G Besio, P Blondeaux, V van Lancker, E Verfaillie and G Vittori (2007): Modeling sand wave characteristics on the Belgian continental shelf and in the Calais-Dover Strait. *J Geophys Res, Vol 112, No C6,* Article Number: C06002.

Clifton, H E (1976): Wave formed sedimentary structures – A conceptual model. *SEPM Special Publication No 24,* pp 126-148.

Coleman, N (1970): Flume studies of the sediment transfer coefficient. *Water Resources Research, Vol 6 No 3,* pp 801-809.

Conley, D C and D L Inman (1992): Field observations of the fluid-granular boundary layer flow under near breaking waves. *J Geophys Res, Vol 97, No C6,* pp 9631-9643.

Conley, D C and D L Inman (1994): Ventilated oscillatory boundary layers. *J Fluid Mech, Vol 273,* pp 261-284.

Conley, D C, S Falchetti, I P Lohman and M Brocchini (2008): The effects of flow stratification by non-cohesive sediment on transport in high-energy wave-driven flows. *J Fluid Mech, Vol 610,* pp 43-67.

Crank, J (1956): *The Mathematics of Diffusion.* Oxford University Press, London.

Csanady, G T (2001): *Sea-Air Interaction: Laws and Mechanisms.* Cambridge University Press, 239pp.

Dallimore, C L, J Imberger and T Ishikawa (2001): Entrainment and turbulence in saline underflow in Lake Ogwara. *J Hydr Eng, ASCE, Vol 127, No 11,* pp 937-948.

Darwin, G H (1962): *The Tides and Kindred Phenomena in the Solar System.* Freeman, San Francisco.

Davis, G A, D J Hanslow, K Hibbert and P Nielsen (1992): Gravity drainage: A new method of beach stabilisation through drainage of the watertable. *Proc 23rd Int Conf Coastal Eng,* Venice, ASCE, pp 1129-1141.

Davis, J P, D J Walker, M Townsend and I R Young (2004): Wave-formed sediment ripples: Transient analysis of ripple spectral development. *J Geophys Res, Vol 109,* C07020.

Dean, R G (1974): Evaluation and development of water wave theories for engineering applications. U S Army Corps of Engineers, Coastal Engineering Research Center, Special Report No 1. 2 volumes.

Dean, R G (1991): Equilibrium beach profiles: Principles and applications. *J Coastal Res, Vol 7, No 1,* pp 53-84.

Dean, R G (2002): *Beach Nourishment: Theory and Practice.* World Scientific, River Edge N J, 399pp.

Dean, R G and R A Dalrymple (1991): *Water Wave Mechanics for Engineers and Scientists.* World Scientific, Singapore, 353pp.

References

Dean, R G and R A Dalrymple (2002): *Coastal Processes with Engineering Applications*. Cambridge University Press, 475pp.

Defant, A (1958): *Ebb and Flow*. University of Michigan Press, Ann Arbor.

Del Valle, R, R Medina and M A Losada (1993): Dependence of coefficient on grain size. *J Waterway Port and Coastal Engineering, Vol 119, No 5*, pp 568-574.

Dibajnia, M and A Watanabe (1998): Transport rate under irregular sheet flow conditions. *Coastal Engineering, Vol 35*, pp 167-183.

Dick J E and J F A Sleath (1991): Velocity and concentrations in oscillatory sheet flow over beds of sediment. *J Fluid Mech, Vol 223*, pp 165-196.

Dick J E and J F A Sleath (1991): Sediment transport in oscillatory sheet flow. *J Geophys Res, Vol 97, No C4*, pp 5745-5758.

Dingemans, M W (1997): *Water Wave Propagation over Uneven Bottoms, Vols I and II*. World Scientific.

Dingler, J R (1974): Wave formed ripples in nearshore sands. PhD thesis, Univ. of California San Diego, 136pp.

Dixen, M, F Hatipoglu, B M Sumer and J Fredsøe (2008): Wave boundary layer over a stone covered bed. *Coastal Engineering, Vol 55, No 1*, pp 1-20.

Dohmen-Janssen, C M and D M Hanes (2002) Sheet flow dynamics under monocromatic non-breaking waves. *J Geophys Res, Vol 107, No C10*, 3149, doi:10.1029/2001JC001045.

Dominey-Howes, D (2007): Geological and historical records of tsunami in Australia. *Marine Geology, Vol 239*, pp 99-123.

Donelan, M A, W M Drennan and K B Katsaros (1997): The air sea momentum flux in conditions of wind sea and swell. *J Phys Oceanogr, Vol 27*, pp 2087-2099.

Donelan, M A, B K Haus, N Reul, W J Plant, M Stiassnie, H C Graber, O B Brown and E S Saltzman (2004): On the limiting aerodynamic roughness of the ocean in very strong winds. *Geophys Res Lett, Vol 31*, L18306, doi:10.1029/2004GL019460.

Dronkers J J (1964): *Tidal Computations*. North Holland Publishing Company.

Dronkers, J (2005): *Dynamics of Coastal Systems*. World Scientific, 519pp.

Dunn, S L (2001): Wave setup in river entrances. PhD thesis, The University of Queensland, 175pp.

Dunn, S L, P Nielsen, P A Madsen and P Evans (2000): Wave setup in river entrances. *Proc 27th Int Conf Coastal Engineering*, Sydney, ASCE, pp 3432-3445.

Dyer, K R (1997): *Estuaries: A Physical Introduction*. Wiley, 195pp.

Dysthe, K B and A Harbitz (1987): Big waves from polar lows. *Tellus, Vol 39A*, pp 500-508.

Einstein, A (1912): In *La Theorie du Rayonnement et les Quanta*, Rapports et Discussions, Solvay Conference, Bruxelles 1911, P Langevin and M de Broglie, Gauthier-Villars, Paris, p 450.

References

Ekman, V W (1905): On the influence of the Earth's rotation on ocean currents. *Arkiv f Matematik, Astronomi & Fysik*, Vol 2 No 11, Stockholm.

Elfrink, B, D M Hanes and B G Ruessink (2006): Parameterization and simulation of near bed orbital velocities under irregular waves in shallow water. *Coastal Engineering, Vol 53, No 11*, pp 915-927

Elgar, S, R T Guza and M H Freilich (1988): Eulerian measurements of horizontal accelerations in shoaling gravity waves. *J Geophys Res, Vol 93, No C8*, pp 9261-9269.

Elgar, S, E L Gallagher and R T Guza (2001): Nearshore sandbar migration. *J Geophys Res, Vol 106, No C3*, pp 11623-11627.

Ellis, K M, C E Binding, D G Bowers, S E Jones and J H Simpson (2008): A model of turbidity maximum maintenance in the Irish Sea. *Estuarine, Coastal and Shelf Science, Vol 76, No 4*, pp 765-774.

Farachi, C and E Foti (2002): Geometry, migration and evolution of small scale bedforms generated by regular and irregular waves. *Coastal Engineering, Vol 47*, pp 35-52.

Feddersen, F, R T Guza, S Elgar and T H C Herbers (1996): Cross shore structure of longshore currents during Duck '94. *Proc 25th Int Conf Coastal Eng*, Orlando, ASCE, pp 3666-3679.

Feddersen, F, R T Guza and S Elgar (2004): Inverse modelling of one-dimensional setup and alongshore current in the nearshore. *J Physical Oceanography, Vol 34*, pp 920-933.

Fenton, J D (1988): The numerical solution of steady water wave problems. *Computers and Geosciences, Vol 14, No 3*, pp 357-368.

Fenton, J D (1990): Nonlinear wave theories. *The Sea, Vol 9*, Editors B Le Mehaute and D M Hanes, Wiley, New York.

Fernando, H J S (1991): Turbulent mixing in stratified fluids. *Ann Rev Fluid Mech, Vol 23*, pp 455-493.

Fisher, H B, E J List, R C Y Koh, J Imberger and N H Brooks (1979): *Mixing in Inland and Coastal Waters*. Academic Press, N Y.

Flick, R E, R T Guza and D L Inman (1981): Elevation and velocity measurements in laboratory shoaling waves. *J Geophys Res, Vol 86, No C5*, pp 4149-4160.

Fredsøe, J (1984): Turbulent boundary layers in wave current motion. *J Hydr Eng, ASCE, Vol 110, Hy 10*, pp 1103-1120.

Fredsøe, J and R Deigaard (1992): *Mechanics of Coastal Sediment Transport*. World Scientific, 369pp.

Fredsøe, J, K H Andersen and B M Sumer (1999): Wave plus current over a ripple covered bed. *Coastal Engineering, Vol 38*, pp 177-221.

Friedman, P D and J Katz (2002): Mean rise rate of droplets in isotropic turbulence. *Physics of Fluids, Vol 14, No 9*, pp 3059-3073.

Fritz, H M, W H Hager and H-E Minor (2004): Near field characteristics of landslide generated impulse waves. *J Waterway, Port, Coastal and Ocean Eng. Vol 130, No 6*, pp 287-302.

References

Gallagher, E L (2003): A note on megaripples in the surf zone. Evidence for their relation to steady flow dunes. *Marine Geology, Vol 193*, pp 171-176.

Gallagher, E L, S Elgar and E B Thornton (1998): Megaripple migration in a natural surf zone. *Nature, Vol 394*, pp 165-168.

Galvin, C J (1970): Finite amplitude shallow water waves of periodically recurring form. Reprint 5-72. Coastal Engineering Research Center, US Army Corps of Engineers, 32pp.

Goda, Y (1988): On the method of selecting design wave height. *Proc 21st Int Conf Coastal Eng, Malaga*, ASCE, pp 899-913.

Gourlay, M R and G Colleter (2005): Wave generated flow on coral reefs — an analysis for 2-dimensional horizontal reef tops with steep faces. *Coastal Engineering, Vol 52, No 4*, pp 353-388.

Graf, W H and M Cellino (2002): Suspension flows in open channels; experimental study. *Hydraulic Res, Vol 40, No 4*, pp 435-447.

Grassmeier, B T and L C van Rijn (1999): Transport of fine sand by current and waves, Part II: Breaking waves over barred profile with ripples. *J Waterway, Port, Coastal and Ocean Eng. Vol 125, No 2*, pp 71-79.

Green, W H and G A Ampt (1911): Studies in soil physics 1. The flow of air and water through soils. *J Agric Sci, Vol IV, No I*, pp 1-24.

Gregg, M C (1987): Diapycnal mixing in the thermocline: A review. *J Geophys Res, Vol 92, C5*, pp 5249-5286.

Grimshaw, R H J, D-H Zhang and K W Chow (2007): Generation of solitary waves by transcritical flow over a step. *J Fluid Mech, Vol 587*, pp 235-254.

Guard P A, and T E Baldock (2007): The influence of seaward boundary conditions on swash zone hydrodynamics. *Coastal Engineering, Vol 54, No 4*, pp 321-331.

Hamm, L and C Migniot (1994): Elements of cohesive sediment deposition, consolidation and erosion. In M B Abbott and A Price ed (1994): *Coastal Estuarial and Harbour Engineers' Reference Book*. E & F N Spon, pp 93-106.

Hancock, M W and N Kobayashi (1994): Wave overtopping and sediment transport over dunes. *Proc 24th Int Conf Coastal Eng*, Kobe, ASCE, pp 2028-2042.

Hanes, D M (1990): The structure of events of intermittent suspension of sand due to shoaling waves, *The Sea Vol 9*, John Wiley & Sons, pp 941-952.

Hanes, D M and A J Bowen (1985): A granular fluid model for steady, intense bedload transport. *J Geophys Res, Vol 90, No C5*, pp 9149-9158.

Hanslow, D J and P Nielsen (1993): Shoreline setup on natural beaches. *J Coastal Res*, Special Issue 15, pp 1-10.

Hanslow, D J, P Nielsen and K Hibbert (1996): Wave setup at river entrances. *Proc 25th Int Conf Coastal Eng*, Orlando, September 1996, ASCE, pp 2244-2257.

Hay, A E and D J Wilson (1994): Rotary sidescan images of nearshore bedform evolution during a storm. *Marine Geology, Vol 119*, pp 57-65.

Hinwood, J, E McLean and M Trevethan (2005): Spring tidal pumping. *Proc 17th Australasian Coastal Eng Conf*, Adelaide, pp 601-606.

Ho, H W (1964): Fall velocity of a sphere in an oscillating fluid. PhD thesis, University of Iowa.

Hogben, N (1990): Long term wave statistics. *The Sea, Vol 9*, Editors B Le Mehaute and D M Hanes, Wiley, New York.

Holthuijsen, L H (2007): *Waves in Oceanic and Coastal Waters*. Cambridge University Press, 387pp.

Horikawa, K (1988): *Nearshore Dynamics and Coastal Processes*. University of Tokyo Press, 522pp.

Horikawa, K, A Watanabe and S Katori (1982): Sediment transport under sheet flow condition. *Proc 18th Int Conf Coastal Eng, Capetown*, ASCE, pp 1335-1352.

Hughes, M G and A S Moseley (2007): Hydrokinematic regions within the swash zone. *Cont Shelf Res, Vol 27*, pp 2000-2013.

Hunt, I A (1959): Design of seawalls and breakwaters. *Proc ASCE, Vol 85*, pp 123-152.

Huppert, H E, J S Turner and M A Hallworth (1995): Sedimentation and entrainment in dense layers of suspended particles stirred by an oscillating grid. *J Fluid Mech, Vol 289*, pp 263-293.

Inman, D L and R A Bagnold (1963): Littoral processes. In N M Hill (ed) *The Sea, Vol III*, Interscience, NY, pp 529-553.

Inman, D L and A J Bowen (1963): Flume experiments on sand transport by waves and currents. *Proc 8th Int Conf Coastal Eng*, ASCE, pp 137-150.

Institution of Engineers Australia (2004): *Coastal Engineering Guidelines for Working with the Australian Coast in an Ecologically Sustainable Way*. ISBN 0-85825-819-6, EA Books, 128pp.

Institution of Engineers Australia (2004): *Guidelines for Responding to the Effects of Climate Change in Coastal and Ocean Engineering*. ISBN 0-85825-750-5, EA Books, 76pp.

Irish, J L, D T Resio and J J Ratcliff (2008): The influence of storm size on hurricane surge. *J Phys Oceanography, Vol 38*, pp 2003-2013.

Ivey, G N and J Imberger (1991): On the nature of turbulence in a stratified fluid. Part I: The energetics of mixing. *J Phys Oceanography, Vol 21*, pp 650-659.

Janssen, T T (2006): *Nonlinear surface waves over topography*. PhD thesis, Delft University of Technology, 208pp.

Janssen, T T, J A Battjes and A R van Dongeren (2003): Long waves induced by short wave groups over a sloping bottom. *J Geophys Res, Vol 108, No C8*, 3252, doi: 10.1029/2002JC001515.

Jenkins, G M and D G Watts (1968): *Spectral Analysis and its Applications*. Holden-Day, 525pp.

Jensen, B L, B M Sumer and J Fredsoe (1989): Turbulent oscillatory boundary layers at high Reynolds numbers. *J Fluid Mech, Vol 206*, pp 265-297.

References

Jobson, H E and W W Sayre (1970): Vertical transfer in open channel flow. *J Hydr Div, ASCE, Vol 96, No Hy3*, pp 703-724.

Jones, I S F and Y Toba (2001): *Wind Stress over the Ocean*. Cambridge University Press.

Jonsson, I G (1980): A new approach to rough turbulent oscillatory boundary layers. *Ocean Eng, Vol 7*, pp 109-152.

Jonsson, I G (1990): Wave-current interactions. *The Sea, Vol 9*, Editors B Le Mehaute and D M Hanes, Wiley, New York.

Justesen, P (1988): Turbulent wave boundary layers. Series Paper 43, ISVA, T U Denmark, 226pp.

Kamphuis, J W (1990): Littoral sediment transport rate. *Proc 22nd Int Conf Coastal Eng, Delft*, ASCE, pp 2402-2415.

Kamphuis, J W (1991): Alongshore sediment transport rate distribution. *Proc Coastal Sediments '91*, ASCE, pp 170-183.

Kang, H-Y and P Nielsen (1994): Watertable overheight due to wave runup on a sandy beach. *Proc 24th Int Conf Coastal Eng, Kobe*, ASCE, pp 2115-2124.

Kaczmarek, L M and R Ostrowski (2002): Modelling intensive near bed sand transport under wave-current flow versus laboratory and field data. *Coastal Engineering, Vol 45, No 1*, pp 1-18.

Kerr, D (2007): Marine energy, *Phil Trans Roy Soc Lond, Vol 365*, pp 971-992.

Keulegan, G H (1966): The mechanism of an arrested saline wedge. In A T Ippen ed, *Estuary and Coastline Hydrodynamics*, McGraw-Hill, pp 546-597.

Kharif, C and E Pelinovsky (2003): Physical mechanisms of the rogue wave phenomenon. *European Journal of Mechanics B/Fluids, Vol 22*, pp 603-634.

King, D B (1991): Studies in oscillatory flow bedload sediment transport. PhD thesis, Univ. of California, San Diego, 184pp.

Kinsman, B (1965): *Wind Waves: Their Generation and Propagation on the Ocean Surface*. Prentice-Hall.

Kirby, J T (1986a): Open boundary-condition in parabolic equation method. *Journal of Waterway Port Coastal and Ocean Engineering-ASCE, Vol 112, No 3*, pp 460-465.

Kirby, J T (1986b). On the gradual reflection of weakly nonlinear stokes waves in regions with varying topography. *Journal of Fluid Mechanics, Vol 162*, pp 187-209.

Kirby, J T (1986c): Higher-order approximations in the parabolic equation method for water-waves. *Journal of Geophysical Research-Oceans, Vol 91, (C1)*, pp 933-952.

Kirby, J T (1986d): Rational approximations in the parabolic equation method for water waves. *Coastal Engineering, Vol 10*, pp 355-378.

Kirby, J T and R A Dalrymple (1983a): A parabolic equation for the combined refraction diffraction of Stokes waves by mildly varying topography. *Journal of Fluid Mechanics, 136(Nov)*, pp 453-466.

Kirby, J T and R A Dalrymple (1983b): The propagation of weakly non-linear wave in the presence of varying depth and currents, *Proc XXth Congress I.A.H.R.*, Moscow.

Kirby, J T and R A Dalrymple (1984): Verification of a parabolic equation for propagation of weakly-nonlinear waves. *Coastal Engineering, Vol 8, No 3*, pp 219-232.

Kirby, J T and R A Dalrymple (1986a). Modelling waves in surfzones and around islands. *Journal of Waterway Port Coastal and Ocean Engineering-ASCE, Vol 112.*

Kirby, J T and R A Dalrymple, (1986b): An approximate model for nonlinear dispersion in monochromatic wave propagation models. *Coastal Engineering, Vol 9*, pp 545-561.

Kirby, J T and R A Dalrymple (1994): Combined refraction/diffraction model REF/DIF1, Version 2.5. Res. Report CACR-94-22, Center for Applied Coastal Research, Univ. of Delaware.

Kjerfve, B (1988): *Hydrodynamics of Estuaries, Vols I, II.* CRC Press.

Klopman G (1994): Vertical structure of the flow due to waves and currents. Laser Doppler flow measurements for waves following or opposing a current. Tech Rep H840.32 Part 2, Delft Hydraulics.

Kobayashi, N and E A Karjadi (1994): Swash dynamics under obliquely incident waves. *Proc 22nd Int Conf Coastal Eng*, Delft, ASCE, pp 2155-2169.

Komar, P D (1998): *Beach Processes and Sedimentation.* Prentice Hall.

Korteweg, D J and G de Vries (1895): On the change of form of long waves advancing in a rectangular channel and on a new type of long, stationary waves. *Phil Mag, Series 5, Vol 39*, pp 422-443.

Kostic, S and G Parker (2003): Physical and numerical modelling of deltaic sedimentation in lakes and reservoirs. *Proc 30th IAHR Conf, Theme C, Vol 1*, pp 413-420.

Kraus, N C, M Larson and D L Kriebel (1991): Evaluation of beach erosion and accretion predictors. *Proc Coastal Sediments '91*, ASCE, pp 572-587.

Krause, D C, W C White, D J W Piper and B C Heezen (1970): Turbidity currents and cable breaks in the western new Britain Trench. *Bull Geol Soc Am, Vol 81*, pp 2153-2160.

Kriebel, D L and R G Dean (1985): Numerical simulation of time-dependent beach profile response. *Coastal Engineering Vol 9, No 3*, pp 221-245.

Landau, L D and E M Lifschitz (1987): *Course of Theoretical Physics, Vol 6: Fluid Mechanics*, 2nd ed, Butterworth Heineman, 539pp.

Le Mehaute, B (1976): *An Introduction to Hydrodynamics and Water Waves.* Springer-Verlag, 313pp.

Lewis, R (1997): *Dispersion in Estuaries and Coastal Waters.* John Wiley, Chichester, UK 312pp.

Liggett, J A (1994): *Fluid Mechanics.* McGraw-Hill, 495pp.

Lighthill, M J (1978): *Waves in Fluids.* Cambridge University Press.

Liu, P F-L, Y S Park and E A Cowen (2007): Boundary layer flow and bed shear stress under a solitary wave. *J Fluid Mech, Vol 574*, pp 449-463.

Löfquist, K E B (1978): Sand ripple growth in an oscillatory-flow tunnel. Tech Paper No 78-5, US Army Corps of Engineers Coastal Engineering Research Center, 101pp.

Löfquist, K E B (1986): Drag on naturally rippled beds under oscillatory flows. Misc Paper CERC-86-13, U S Army Corps of Engineers, 121pp.

Longuet-Higgins, M S (1953): Mass transport in water waves. *Phil Trans Roy Soc Lond, Vol A 245*, pp 535-581.

Longuet-Higgins, M S (1956): The mechanics of the boundary layer near the bottom in a progressive wave. *Proc 6th Int Conf Coastal Eng*, Miami, ASCE, pp 184-193.

Longuet-Higgins, M S (1970): Longshore currents generated by obliquely incident seawaves. *J Geophys Res, Vol 75, No 33*, pp 6778-6801.

Longuet-Higgins, M S (2005): On wave setup in shoaling water with a rough sea bed. *J Fluid Mech, Vol 527*, pp 217-234

Longuet-Higgins, M S and R W Stewart (1962): Radiation stress and mass transport in gravity waves with application to surf beats. *J Fluid Mech, Vol 13*, pp 481-504.

Longuet-Higgins, M S and R W Stewart (1964): Radiation stresses in water waves; a physical discussion with applications. *Deep Sea Research, Vol 11*, pp 529-562.

Lumley, J L (1962): The mathematical nature of the problem of relating Lagrangian and Eulerian functions in turbulence. In: *Mechanique de la turbulence, Colloq Int CNRS Marseille*, CNRS Paris, pp 17-26.

Lundgren, H (1963): Wave thrust and wave energy level. *Proc 10th Congr Int Ass Hydraulic Res*, London, pp 147-151.

MacMahan, J H, E B Thornton and J H M Reniers (2006): Rip current review. *Coastal Engineering, Vol 53*, pp 191-208.

Madsen, P A and D R Fuhrman (2008): Runup of tsunami and long waves in terms of surf similarity. *Coastal Eng, Vol 55*, pp 208-223.

Magnaudet, J and L Thais (1995): Orbital rotational motion and turbulence below laboratory wind waves. *J Geophysical Res, Vol 100, No C1*, pp 757-771.

Mangor, K (2001): *Shoreline Management Guidelines*. Obtainable from DHI Water and Environment (price DKK 250), e-mail dhi@dhi.dk, 232pp.

Marieu, V, P Bonneton, D L Foster and F Ardhuin (2008): Modelling of vortex ripple morphodynamics. *J Geophys Res, Vol 113, C09007*, doi: 10.1029/2007JC004659.

Martin, C S (1970): Effect of a porous sand bed on incipient sediment motion. *Water Resources Res, Vol 6, No 4*, pp 1162-1174.

Massel, S R (1996): *Ocean Surface Waves: Their Physics and Prediction*. World Scientific. Singapore, 499pp.

References

Masselink, G and M G Hughes (2003). *Introduction to Coastal Processes and Geomorphology*. Edward Arnold, 354pp.

Mathiesen, M, Y Goda, P J Hawkes, E Mansard, M J Martin, E Peltier, E F Thompson and G van Vledder (1994): Recommended practice for extreme wave analysis. *J Hydr Res, Vol 32, No 6*, pp 803-814.

Maxey, M R (1990): On the advection of spherical and non-spherical particles in a non-uniform flow. *Phil Trans Roy Soc Lond, Vol 333*, pp 289-307.

Maxey, M R and S Corrsin (1986): Gravitational settling of aerosol particles in randomly orientated circular flow fields. *J Atmospherical Sci, Vol 43*, pp 1112-1134.

McAnally, W H, C Friedrichs, D Hamilton, E Hayter, P Shresthar, H Rodriguez, A Sheremet and A Teeter (2007): Management of fluid mud in estuaries, bays and lakes. I: Present state of understanding on character and behaviour. *J Hydraulic Eng, Vol 133 No 1*, pp 9-22.

McAnally, W H, A Teeter, D Schoelhamer, C Friedrichs, D Hamilton, E Hayter, P Shresthar, H Rodriguez, A Sheremet and R Kirby (2007): Management of fluid mud in estuaries, bays and lakes. II: Measurement modelling and management. *J Hydraulic Eng, Vol 133 No 1*, pp 23-38.

Mei, C C (1989): *The Applied Dynamics of Ocean Surface Waves*. World Scientific.

Meyer, R E and A D Taylor (1972): Runup on beaches. In *Waves on Beaches and the Resulting Sediment Transport*. Academic Press, pp 357-411.

Miles, J W (1981): The Korteweg-de Vries equation: A historical essay. *J Fluid Mech, Vol 106*, pp 131-147.

Mitchener, H and H Torfs (1996): Erosion of mud/sand mixtures. *Coastal Engineering, Vol 29, No 1*, pp 1-25.

Murray, S P (1970): Settling velocity and vertical diffusion of particles in turbulent water. *J Geophys Res, Vol 75, No 9*, pp 1647-1654.

Muste, M, K You, I Fujita and R Ettema (2005): Two-phase versus mixed-flow perspective on suspended sediment transport in turbulent channel flows. *Water Resources Res, Vol 41, W10402*, doi 10.1029/2004WR003595.

Nakato, T, F A Locher, J R Glover and J F Kennedy (1977): Wave entrainment of sediment from rippled beds. *Proc ASCE, Vol 103, No WW1*, pp 83-100.

Nelson, R (1997): Height limits in top down and bottom up wave environments. *Coastal Engineering, Vol 32*, pp 247-254.

Nielsen, P (1979): Some basic concepts of wave sediment transport. PhD thesis, Series Paper 20, ISVA, Tech Univ. Denmark, 160pp.

Nielsen, P (1983): Entrainment and distribution of different sand sizes under water waves. *J Sed Pet, Vol 53, No 2*, pp 423-428.

Nielsen, P (1984): Field measurements of time averaged suspended sediment concentrations. *Coastal Engineering, Vol 8*, pp 51-72.

Nielsen, P (1986): Local approximations: A new way of dealing with irregular waves. *Proc 20th Int Conf Coastal Eng*, Taipei, ASCE, pp 633-646.

References

Nielsen, P (1988): Three simple models of wave sediment transport, *Coastal Engineering, Vol 12*, pp 43-62.

Nielsen, P (1989): Analysis of natural waves by local approximations. *J Waterway, Port Coastal and Ocean Eng*, ASCE, *Vol 115, No 3*, pp 384-396.

Nielsen, P (1990): Tidal dynamics of the watertable in beaches. *Water Resources Research, Vol 26 No 9*, pp 2127-2135.

Nielsen, P (1991): Combined convection and diffusion: A new framework for suspended sediment modelling. *Proc Coastal Sediments '91*, ASCE, pp 418-431.

Nielsen, P (1992): *Coastal Bottom Boundary Layers and Sediment Transport.* World Scientific, Singapore, 324pp.

Nielsen, P (1992): Average velocity of non neutral particles in turbulence. *Proc 11th Australasian Fluid Mech Conf*, Hobart, I E Aust, pp 179-181.

Nielsen, P (1993): Turbulence effects of the settling of suspended particles. *J Sed Petrology, Vol 63, No 5*, pp 835-838.

Nielsen, P (1998): Coastal groundwater dynamics. *Proc Coastal Dynamics '97*, Plymouth, ASCE, New York, pp 546-555.

Nielsen, P (1999): Groundwater dynamics and salinity in coastal barriers. *J Coastal Res, Vol 15, No 3*, pp 732-740.

Nielsen, P (2002): Shear stress and sediment transport calculations for swash zone modeling. *Coastal Engineering, Vol 45, No 1*, pp 53-60.

Nielsen, P (2006): Sheet flow sediment transport under waves with acceleration skewness and boundary layer streaming. *Coastal Engineering Vol 53, No 9*, pp 749-758.

Nielsen, P and D J Hanslow (1991): Wave runup distributions on natural beaches. *J Coastal Res, Vol 7, No 4*, pp 1139-1152.

Nielsen, P and Z-J You (1996): Eulerian mean velocities under non-breaking waves over horizontal bottoms. *Proc 25th Int Conf Coastal Eng*, Orlando, ASCE, pp 4066-4078.

Nielsen, P and I L Turner (2000): Watertable waves and water exchange in beaches. *Proc 27th Int Conf Coastal Engineering*, Sydney, ASCE, pp 2356-2363.

Nielsen, P, R W Brander and M G Hughes (2001): Rip currents: Observations of hydraulic gradients, friction factors and wave pump efficiency. *Proc Coastal Dynamics '01*, Lund, Sweden, ASCE, pp 483-492.

Nielsen, P, S Robert, B Moeller-Christiansen and P Oliva (2001): Infiltration effects on sediment mobility under waves. *Coastal Engineering, Vol 42, No 2*, pp 105-114.

Nielsen, P, K U van der Wal and L Gillan (2002): Vertical fluxes of sediment in oscillatory sheet-flow. *Coastal Engineering, Vol 45, No 1*, pp 61-68.

Nielsen P and D P Callaghan (2003): Shear stress and sediment transport calculations for sheet flow under waves. *Coastal Engineering, Vol 47, No 3*, pp 347-354.

References

Nielsen, P and I A L Teakle (2004): Finite mixing length modelling of the diffusion of fluid momentum and particles in turbulence. *Physics of Fluids, Vol 16, No 7*, pp 2342-2348.

Nielsen, P, P A Guard, D P Callaghan and T E Baldock (2008): Observations of wave pump efficiency. *Coastal Engineering, Vol 55, No 1*, pp 69-72.

Nielsen, P, S de Brye, D P Callaghan and P A Guard (2008): Transient dynamics of storm surges and other forced long waves, *Coastal Engineering*.

Nordstrom, K F and C T Roman (1996): *Estuarine Shores*. Wiley, 479pp.

Open University (1999): *Waves, Tides and Shallow-Water Processes*. 2nd ed. Butterworth Heineman, 227pp.

O'Donoghue, T, M Li, J Malarkey, S Pan, A G Davies and B A O'Connor (2004): Numerical and experimental study of wave generated sheet flow. *Proc 29th ICCE*, Lisbon, World Scientific, pp 1690-1702.

Parker, G (1978): Self formed straight rivers with equilibrium banks and mobile bed. Part 1. The sand-silt river. *J Fluid Mech, Vol 89, Part 1*, pp 109-125.

Parker, G (1982): Conditions for the ignition of catastrophically erosive turbidity currents. *Mar Geol*, Vol 46, pp 307-327.

Parker, G, Y Fukushima and H M Pantin (1986): Self-accelerating turbidity currents. *J Fluid Mech, Vol 171*, pp 145-181.

Peltier, W R and C P Caulfield (2003): Mixing efficiency in stratified shear flows. *Ann Rev Fluid Mech, Vol 35*, pp 135-168.

Phillips, O M (1966): *The Dynamics of the Upper Ocean*. Cambridge University Press.

Philip, J R (1973): Periodic non-linear diffusion: An integral relation and its physical consequences. *Aust J Physics, Vol 26*, pp 513-519.

Pugh, D (2004): *Changing Sea Levels*. Cambridge University Press, 265pp.

Queensland Government (2001): Queensland climate change and community vulnerability to tropical cyclones: Ocean hazards assessment stage 1, 318pp. ISBN 0 7345 1765 3.

Raichlen, F (1966): Harbour resonance, in Ippen, A T: *Estuary and Coastline Hydrodynamics*, McGraw-Hill, pp 281-340.

Readers Digest (1983): *Guide to the Australian Coast*. Surry Hills, NSW, Australia, 479pp.

Ribberink J S and Z Chen (1993): Sediment transport of fine sand under asymmetric oscillatory flow. *Delft Hydrualics Report H840.20, part VII*.

Ribberink, J S and A A Al-Salem (1994): Sediment transport in oscillatory boundary layers in cases of rippled beds and sheet flow. *J Geophys Res, Vol 99, No C6*, pp 12707-12727.

Ribberink, J S, I Katapodi, K A H Ramadan, R Koelewijn and S Longo (1994): Sediment transport under non-linear waves and currents. *Proc 24th Int Conf Coastal Eng*, Kobe ASCE, pp 2527-2541.

Ribberink, J S, C M Dohmen-Janssen, D M Hanes, S R McLean and C Vincent (2000): Near bed sand transport mechanisms under waves. *Proc 27th Int Conf Coastal Eng, Sydney*, ASCE, New York, pp 3263-3276.

References

Richardson, J F and W N Zaki (1954): Sedimentation and fluidization, Part 1. *Trans Inst Chem Eng, Vol 32*, pp 35-53.

Riedel, H P (1972): Direct measurement of bed shear stress under waves. PhD thesis, Queens University, Kingston Ontario, 142pp.

Robinson, C, B Gibbes and L Li (2006): Driving mechanisms for groundwater flow and salt transport in a subterranean estuary. *Geophysical Research Letters, Vol 33*, L03402.

Roelvink J A (2006): Coastal morphodynamics evolution techniques. *Coastal Engineering, Vol 53*, pp 277-287.

Roelvink, J A and I Broeker (1993): Cross-shore profile models. *Coastal Engineering, Vol 21*, pp 163-191.

Rouse, H (1937): Modern conceptions of the mechanics of turbulence. *Trans ASCE, Vol 102*, pp 463-543.

Rubin, D M and R M Hunter (1987): Bedform alignment in directionally varying flows. *Science, Vol 237*, pp 276-278.

Rutherford, J C (1994): *River Mixing*. John Wiley, Chichester, USA, 347pp.

Sato, S, M B Kabiling and H Suzuki (1992): Prediction of near bottom velocity history by a non-linear dispersive wave model. *Coastal Eng in Japan, Vol 35, No 1*, pp 67-82.

Sato S (1996): Numerical simulation of 1993 southwest Hokkaido earthquake tsunami around Okushiri Island. *J Waterway, Port and Coastal Engineering, ASCE, Vol 122, No 5*, pp 209-215.

Savanije, H H G (2005): *Salinity and Tides in Alluvial Estuaries*. Elsevier, 194pp.

Sawamoto, M and T Yamashita (1986): Sediment transport rate due to wave action. *J Hydroscience and Hydraulic Eng, Vol 4, No 1*, pp 1-15.

She, K, L Trim and D Pope (2005): Fall velocities of natural sediment particles: As simple mathematical presentation of the fall velocity law. *J Hydraulic Res, Vol 43, No 2*, pp 189-195.

Sherman, D J and E J Farrell (2008): Aerodynamic roughness lengths over movable beds: Comparison of wind tunnel and field data. *J Geophys Res, Vol 113, F02S08*, doi:10.1029/2007/JF000784.

Shields, A (1936): Anwendung der Ähnlichkeitsmechanik und Turbulenzforschung auf die Geschiebe-bewegung. *Mitt Preuss Versuchsanstalt für Wasserbau und Schiffbau, No 26*, Berlin.

Short, A D Ed (1999): *Handbook of Beach and Shore Face Morphodynamics*. Wiley.

Shuto, N (1985): The Nihonkai-Chubo earthquake tsunami on the north Akita coast. *Coastal Eng in Japan, Vol 28*, pp 255-264.

Simons, R R, D Myrhaug, L Thais, G Chapalain, L-E Holmedal and R MacIver (2000): Bed friction in combined wave current flows. *Proc 27th Int Conf Coastal Eng, ASCE*, pp 216-226.

Sistermans, P G J (2002): Graded sediment transport by non-breaking waves and a current. PhD thesis, Delft Univ Tech, 205pp.

References

Sleath J F A (1970): Measurements close to the bed in a wave tank. *J Fluid Mech, Vol 42*, pp 111-123.

Sleath, J F A (1984): *Sea Bed Mechanics*, Wiley, 335pp.

Sleath, J F A (1987): Turbulent oscillatory flow over rough beds. *J Fluid Mech, Vol 182*, pp 369-409.

Sleath, J F A (1999): Conditions for plug formation in oscillatory flow. *Cont Shelf Res, Vol 19*, pp 1643-1664.

Snyder, W H and J L Lumley (1971): Some measurements of particle velocity autocorrelation functions in a turbulent flow. *J Fluid Mech, Vol 48*, pp 41-71.

Soulsby, R L (1997): *Dynamics of Marine Sands: A Manual for Practical Applications*. Thomas Telford, London.

St Laurent, L and H Simmons (2006): Estimates of power consumed by mixing in the ocean interior. *J Clim, Vol 19*, pp 4877-4890.

Stockdon, H F, R A Holman, P A Howd and A H Sallenger Jr (2006): Empirical parameterization of setup, swash and runup. *Coastal Eng, Vol 53, No 7*, pp 573-588.

Stokes, G G (1846): Report on recent researches in hydrodynamics. Rep 16th meeting Br Ass Adv Sci.

Stokes, G G (1847): On the theory of oscillatory waves. *Trans Cambridge Phil Soc, Vol 8*, pp 441-473.

Sumer, B M, A Kozakiewicz, J Fredsoe and R Deigaard (1999): Velocity and concentration profiles in sheet flow layer of mobile bed. *J Hydraulic Eng, Vol 122, No 10*, pp 549-558.

Svendsen, I A (1984): Wave heights and setup in he surf zone. *Coastal Engineering, Vol 8 No 4*, pp 303-329.

Svendsen, I A (2005): *Introduction to Nearshore Hydrodynamics. Adv Series in Ocean Eng Vol 24*. World Scientific.

Svendsen, I A and I G Jonsson (1976): *Hydrodynamics of Coastal Regions*. Den Private Ingenioerfond, Lyngby, 282pp.

Synolakis, C E and E N Bernard (2006): Tsunami science before and beyond Boxing Day 2004. *Phil Trans Roy Soc Lond, Vol A 364*, pp 2231-2265.

Taylor, G I (1921): Diffusion by continuous movements. *London Mathematical Society, Vol 20*, pp 196-212.

Taylor, G I (1953): Dispersion of soluble matter in solvent flowing slowly through a tube. *Proc Roy Soc Lond, Series A, Vol 219*, pp 186-202.

Taylor, G I (1954): The dispersion of matter in turbulent flow through a pipe. *Proc Roy Soc Lond, Series A, Vol 223*, pp 446-468.

Teakle, I A L and P Nielsen (2004): Modelling suspended sediment profiles under waves: A finite mixing length theory. *Proc 29 ICCE*, Lisbon, World Scientific, pp 1780-1792.

Teisson, C, M Ockenden, P Le Hir, C Kranenburg and L Hamm (1993): Cohesive sediment transport processes. *Coastal Engineering, Vol 21*, pp 126-162.

References

Testic, F Y, S I Voropayev and H J S Fernando (2005): Adjustment of sand ripples under changing water waves. *Phys Fluids, Vol 17*, 072104.

Terrile, E, A J H M Reniers, M J F Stive, M Tromp and J Verhagen (2006): Incipient motion of coarse particles under regular shoaling waves. *Coastal Engineering, Vol 53, No 1*, pp 81-92.

Thorpe, S A (1973): Measurements on instability and turbulence in a stratified shear flow. *J Fluid Mech, Vol 73*, pp 731-751.

Tinti, S and E Bortolucci (2000): Analytical investigation on tsunami generated by submarine slides. *Annali di Geofisica, Vol 43, No 3*, pp 519-536.

Tinti, S, E Bortolucci and C Chiavettieri (2001): Tsunami excitation by submarine slides in shallow water approximation. *Pure and Applied Geophysics, Vol 158*, pp 759-797.

Tomkins, M R, P Nielsen and M G Hughes (2003): Selective entrainment of sediment graded by size and density under waves. *Journal of Sedimentary Research, 73(6)*, pp 906-911.

Tooby, P F, G L Wick and J D Isacs (1977): The motion of a small sphere in a rotating velocity field: A possible mechanism for suspending particles in turbulence. *J Geophysical Res, Vol 82, No 15C*, pp 2096-2100.

The Open University (1999): *Waves, Tides and Shallow Water Processes*. Butterworth Heineman, 227pp.

Thorne, P D, A G Davies and P S Bell (2009): Observation and analysis of sediment diffusivity profiles over sandy rippled beds under waves. *J Geophys Res, Vol 114*, C02023, doi: 10.1029/2008JC0044944.

Traikovsky, P, D E Hay, J D Irish and J F Lynch (1999): Geometry, migration and evolution of wave orbital ripples at LEO-15. *J Geophys Res, Vol 104, No C1*, pp 1505-1524.

Traikovsky, P (2007): Observations of wave orbital scale ripples and a non-equilibrium time dependent model. *J Geophy Res, Vol 112*, C06026.

Trowbridge J and O S Madsen (1984): Turbulent wave boundary layers: 1 Model formulation and first order solution. *J Geophys Res, Vol 89, No C5*, pp 7989-7997.

Tsuruya, H, S Nakano and H Ichinohe (1984): Experimental study on the deformation of runup of tsunami in shallow water — Case study of the tsunami caused by the 1983 Nihonkai Chubu earthquake. *Proc 31st Japanese Conf Coastal Eng*, pp 237-241.

Turner, I L (1993): Water table outcropping on macro-tidal beaches. *Marine Geology, Vol 115*, pp 227-238.

Turner, I L and S P Leatherman (1997): Beach dewatering as a soft engineering solution to coastal erosion. *J Coastal Res, Vol 13, No 4*, pp 1050-1063.

Turner, I L and P Nielsen (1997): Rapid watertable fluctuations: Implications for swash zone sediment mobility. *Coastal Engineering, Vol 32*, pp 45-59.

Turner J S (1973): *Buoyancy Effects in Fluids*. Cambridge University Press, 367pp.

U S Army Corps of Engineers (1984): Shore Protection Manual.

References

Ursell, F (1952): Edge waves on a sloping beach. *Proc Roy Soc Lond, Vol A214*, pp 97-97.

Van Doorn, Th (1982): Experimenteel onderzoek naar het snelheidsveld in de turbulente bodemgrenslaagin een oscillernde stroming in een golftunnel. Delft Hydraulics Laboratory, Report M1562 1a.

Van Rijn, L C (1984a): Sediment pickup functions. *J Hydraulic Eng, Vol 110, No 10*, ASCE, pp 1494-1502.

Van Rijn, L C (1984b): Sediment transport Part II: Suspended load transport. *J Hydraulic Eng, Vol 110, No 10*, ASCE, pp 1613-1641.

Van Rijn, L C and F J Havinga (1995): Transport of fine sands by current and waves. Part II, *J Waterway Port Coastal and Ocean Engineering. Vol 121, No 2*, pp 123-133.

Van Rijn L C (2007): Unified view of sediment transport by currents and waves. I: Initiation of motion, bed roughness, and bed-load transport. *J Hydraulic Eng*, ASCE, *Vol 133, No 6*, pp 649-667.

Van Rijn L C (2007): Unified view of sediment transport by currents and waves. II: Suspended transport. *J Hydraulic Eng*, ASCE, *Vol 133, No 6*, pp 668-689.

Van Rijn L C (2007): Unified view of sediment transport by currents and waves. III: Graded beds. *J Hydraulic Eng*, ASCE, *Vol 133, No 7*, pp 761-775.

Van Rijn L C, D J R Walstra and M van Ormondt (2007): Unified view of sediment transport by currents and waves. IV: Application of morphodynamic model. *J Hydraulic Eng*, ASCE, *Vol 133, No 7*, pp 776-793.

Villaret, C and G Perrier (1992): Transport of fine sand by combined waves and current: An experimental study. Technical Note, Electricite de France, 81pp.

Watanabe, A and S Sato (2004): A sheet flow transport rate formula asymmetric forward leaning waves and currents. *Proc 29th ICCE*, Lisbon, World Scientific, pp 1703-1714.

Weir F M, M G Hughes and T E Baldock (2006): Beach face and berm morphodynamics fronting a coastal lagoon, *Geomorphology, Vol 82, No 3-4*, pp 331-346.

Wijnberg, K M (1996): On the systematic offshore decay of breaker bars. *Proc 25th ICCE*, Orlando, ASCE, pp 3600-3613.

Wilson, R E (1988): Dynamics of partially mixed estuaries. In Kjerfve, B ed: *Hydrodynamics of Estuaries, Vol I*, CRC Press, pp 1-15.

Winterwerp, J C (1998): A simple model for turbulence induced flocculation of cohesive sediments. *J Hydraulic Res, Vol 36, No 3*, pp 309-326.

Winterwerp, J C and W G M van Kesteren (2004): *Introduction to the Physics of Cohesive Sediment Dynamics in the Marine Environment. Developments in Sedimentology, Vol 56*, Elsevier, 466pp, ISBN 0444515534.

Wood, I R, R G Bell and D L Wilkinson (1993): *Ocean Disposal of Wastewater*. World Scientific, Singapore.

Wolanski, E (2007): *Estuarine Ecohydrology*. Elsevier, 157pp.

Wright, L D and A D Short (1984): Morphodynamic variability of beaches and surf zones, a synthesis. *Marine Geology, Vol 56*, pp 92-118.

References

Wüest, A and A Lorke (2003): Small-scale hydrodynamics in lakes. *Ann Rev Fluid Mech, Vol 35*, pp 373-412.

You, Z-J and D Lord (2005): Influence of the El Nino southern oscillation on storm severity along the NSW coast. *Proc 17th Australasian Coastal Eng Conf*, Adelaide, pp 389-392.

Young, I R and L A Verhagen (1996): The growth of fetch limited waves in finite depth: Part 1. Total energy and peak frequency. *Coastal Engineering, Vol 29*, pp 47-78.

Young, I R and L A Verhagen (1996): The growth of fetch limited waves in finite depth: Part 2. Spectral evolution. *Coastal Engineering, Vol 29*, pp 79-100.

Young, I R, L A Verhagen and S K Khatri (1996): The growth of fetch limited waves in finite depth: Part 3. Directional spectra. *Coastal Engineering, Vol 29*, pp 101-122.

Zala Flores, N and J F A Sleath (1998): Mobile layer in oscillatory sheet flow. *J Geophys Res, Vol 103, No C6*, pp 12,783-12,793.

Index